Planning and Scheduling for Maritime Container Yards

Wenkai Li · Yong Wu · Mark Goh

Planning and Scheduling for Maritime Container Yards

Supporting and Facilitating the Global Supply Network

Springer

Wenkai Li
Graduate School of International Management
International University of Japan
Niigata
Japan

Mark Goh
School of Business IT and Logistics, Platform Technologies Research Institute
RMIT University
Melbourne
Australia

Yong Wu
Department of International Business and Asian Studies
Griffith University
Southport, QLD
Australia

ISBN 978-3-319-17024-4 ISBN 978-3-319-17025-1 (eBook)
DOI 10.1007/978-3-319-17025-1

Library of Congress Control Number: 2015935209

Springer Cham Heidelberg New York Dordrecht London

Printed on acid-free paper

Springer International Publishing AG Switzerland is part of Springer Science+Business Media (www.springer.com)

Preface

This book contains six chapters that can be used on a self-contained basis as teaching cases in an undergraduate or specialist class setting on OR techniques applied to maritime container operations or for port operations. Instructors and students can look forward to an insightful presentation of the challenges facing the maritime supply chains and container port logistics service providers in Asia and their innovative responses to these challenges through the real-world cases presented in this book.

Chapter 1 presents a candid look at the maritime trade logistics and supply chain landscape by examining the trend of globalization and port development within Asia, and its impact on container port business activities and the environment.

Chapter 2 explores the provision of container yard and quay services in detail. It examines how a leading port manages the port efficiently by joint optimization of the quay crane, yard crane, and prime mover, each of which is supported by extensive literature review.

Chapter 3 highlights how container terminal operations are bottlenecked by the yard cranes. So, this chapter develops a discrete time model and its associated algorithms for rapid yard crane scheduling, based on realistic operational constraints such as inter-crane interference, fixed separation distances, and simultaneous storage/retrievals in the yard.

Chapter 4 then goes on to study the continuous time model for multiple yard crane scheduling with additional practical constraints. This chapter specifically looks at the impact of last minute container job insertions and shows how the container yard problem can be solved quickly and robustly in polynomial time. This application is extremely useful for transshipment operations where there are numerous occasions of last minute destination port shifts.

Chapter 5 examines the role of heuristics in yard crane scheduling. We look specifically at two underlying concepts for our heuristics development in the case of multiple yard crane scheduling, namely that of the concurrent and divide and conquer approaches. We then provide a hybrid of the two approaches to yield the clustering-reassigning heuristic which is a better candidate for real-time scheduling. Again, this heuristic has practical industry value as container ports attempt to reduce their time taken for creating loading plans in their bid to be competitive vis-à-vis the other ports nearby.

Chapter 6 revisits the frequent approach used in container scheduling by port operators and that is simulation. Specifically, we apply discrete time simulation to validate and assess the proposed hybrid heuristic in Chap. 5. Further, we deploy a simulation model on a mega transshipment container terminal in Asia and investigate the performance of the heuristic under various conditions such as different routing policies, equipment ratios, and so on.

We trust that you will enjoy reading the book and using the techniques as much as we have enjoyed writing them for you.

The first author would like to thank his family members (father Li Kehua, mother Chen Xianfeng, sister Li Xiaoyan, wife Duan Xiaohong, daughter Li Xiaoxue, and son Li Houzhe) for their patience and perseverance during the many months of writing. Without their support and encouragement this monograph would not have been possible.

Wenkai Li
Yong Wu
Mark Goh

Contents

1 **Introduction** 1
References 4

2 **Maritime Terminal Operational Problems** 5
2.1 Container, Container Vessels, and Configuration of a Container Terminal 5
2.1.1 Quay Area 8
2.1.2 Transport Area 10
2.1.3 Yard Area 11
2.1.4 Practical Example of CT: PSA Singapore Terminals 13
2.2 Decision Problems 14
2.3 Literature Review 16
2.3.1 Quay Area 16
2.3.2 Transport Area 19
2.3.3 Yard Area 20
2.3.4 Integration of Decision Problems 23
References 25

3 **Discrete Time Model and Algorithms for Container Yard Crane Scheduling** 31
3.1 Introduction 31
3.2 Literature Review 34
3.2.1 Some Approaches to YC Work Schedules 34
3.3 The Model (DMIP1) 35
3.3.1 Assumptions 36
3.3.2 Objective Function 36
3.3.3 Constraints 37
3.3.4 Results of DMIP1 40
3.4 Model Integrated with Heuristics (DMIP2) 42
3.4.1 Results of DMIP2 43

3.5 Rolling-Horizon Algorithm for DMIP2 (DMIP3) 44
3.5.1 Constraints and Sets Used in the Rolling-Horizon Algorithm 46
3.5.2 Flow Diagram of Rolling-Horizon Algorithm 46
3.5.3 Results of DMIP3 48
3.6 Influence of Weights in Objective Function 48
3.7 Conclusion 49
Appendix 1: Definition of Sets and Parameters 50
References 51

4 A Continuous-Time Model for Multiple Yard Crane Scheduling with Last-Minute Job Arrivals 53
4.1 Introduction 53
4.1.1 Background 53
4.1.2 Literature Review 54
4.1.3 Contributions 56
4.2 Continuous-Time Model (CMIP1) 58
4.2.1 Objective Function 59
4.2.2 Constraints 59
4.2.3 Assumptions 66
4.2.4 Results of CMIP1 67
4.3 Algorithms Integrated with the Continuous-Time Model 70
4.3.1 Heuristics 70
4.3.2 Rolling-Horizon Algorithm (RHA) 72
4.3.3 Results of the Continuous-Time Model Integrated with Heuristics and RHA 73
4.4 Last-Minute Job Arrivals 76
4.5 Discussions 77
4.5.1 The Best Performance Measure 77
4.5.2 Influence of Objective Weights 78
4.6 Conclusions 78
Appendix 1: Definition of Sets and Parameters 78
Appendix 2: Constraints Used in CMIP2 80
References 83

5 Heuristic Approach 85
5.1 Concurrent Scheduling 85
5.2 Divide and Conquer 87
5.3 Clustering-Reassigning Heuristic 88
5.3.1 The Algorithm 89
5.3.2 Comparison with the Continuous MILP Model 91
5.4 Conclusion 94
References 94

6 Validation via Simulation 97
6.1 Simulation at the Block Level 97
6.2 Simulation at the Terminal Level 99
6.2.1 The Simulation Model 100
6.2.2 The Results 104
6.3 Conclusion 107
References 108

Index 109

Chapter 1
Introduction

This chapter presents an outlook of maritime container terminal by examining the trend of globalization and port development within Asia and the impact on container port business activities and the environment. The importance of investment in Operations Research (OR) techniques on improving the productivities of maritime container terminals is highlighted.

In a globalized world, ports form an integral and critical part of the international trade network, as they provide a highly cost-effective movement of goods across markets. As critical international logistics facilities, these nodes provide a vital link to hinterland production points of a country or landlocked geography, and as gateways to consumption markets elsewhere. As such, all attention of the shipping industry and its allied industries has been focused on port throughput and development. Indeed, there has been much emphasis on port development as governments recognize that trade through ports contributes to a significant proportion of GDP. The number of containers moved serves as a bellwether of the country's economic health and the region's supply chain connectivity, as attested by Zhang et al. (2005). Zhang et al. (2005) work suggests a certain degree of correlation in this instance. However, we leave the details to the interested reader. We instead focus our attention on the core activity of the port, namely, the container activities of movement and storage.

A container port comprises a container terminal and the container yard. From a network perspective, the container terminal forms the node of a network and the yard is the storage point. According to Wikipedia (Wiki_Container terminal), a container terminal is the key logistics facility where cargo containers are moved or transshipped between different transport vehicles, for onward transportation either to its hinterland or for another seabound journey. This transshipment may be between container ships and land vehicles, for example, trains or trucks, in which case the terminal is described as a maritime container terminal. Alternatively, the transshipment may be between land vehicles, typically between train and truck, in which case the terminal is described as an inland container terminal.

W. Li et al., *Planning and Scheduling for Maritime Container Yards*,
DOI 10.1007/978-3-319-17025-1_1

From its humble beginnings in the 1950s, the maritime container terminal has come a long way. Maritime container terminals are now a common feature in many countries with a coastline. In addition, these terminals as part of the port are growing larger in size, and embracing newer technologies and infrastructure. The functions remain the same though, and that in container terminals usually provide storage facilities for both loaded and empty containers. Loaded containers are stored for relatively shorter periods of time, so as not to incur higher than necessary demurrage and other costs. Whilst waiting for onward transportation, in some of the container ports, the laden containers may have to be stored for longer periods awaiting their next use. The operational reasons for this are as follows: the next ship is not ready, the container is not cleared for the next shipment or the shipper wants to optimize the free demurrage time afforded by the terminal operator. In such cases, these containers are normally stacked for storage, in a three-high rectangular block formation, and the resulting stores are then commonly known as container stacks. It is only in Asia that we see the prevalence of container stacks getting higher and higher beyond the three-high limitation. Storing containers higher signals better use of land at the coastline, which also faces increasing competition from real estate developers who appreciate the premium on seafront developments. Hence, there is a need to constantly look for optimizing solutions to improve the efficiency of storage and the movement of these container boxes.

In recent years, methodological advances regarding container terminal operations have improved considerably, particularly in the areas of container operations notably in areas such as berthing, stacking, retrieval of containers, and assigning slots for the container boxes. These improvements in productivity have been made possible using OR, technology using wireless platforms, and modern equipment handling. For a detailed description and a comprehensive list of the literature, the interested reader is referred to the extant operations research literature (Steenken et al. 2004; Stahlbock et al. 2007) and Chap. 2 of this book.

In Asia, there are good reasons for undertaking the investment in OR, optimization, technology equipping, and better material handling infrastructure. From Table 1.1, we note that Asian ports rank highly among the world's top 20 container ports. The volume of containers arriving at and leaving each of these ports has generally been growing for the past decade. With growing container traffic due to faster globalization and the need for faster speed to market to meet increasing consumer demand in Asia, there will continue to be rigorous demand for OR driven productivity improvement in port related applications. Port operators are only too aware that they have to be more efficient in their operations so as to encourage more liners to call at their ports and to reduce any unnecessary dwell time at the ports for operational and security reasons. Given the unique geography of Asia and the intense competition of the container ports (which can be sometimes located close to each other such as Singapore and Tanjung Pelepas Malaysia), the challenge is on how to maximize the throughput of the laden and unladen containers particularly

Table 1.1 World's top 20 busiest container ports

Rank	Port	Country/region	2012	2011	2010
1	Shanghai	China	32,530	31,740	29,069
2	Singapore	Singapore	31,650	29,940	28,431
3	Hong Kong	China	23,120	24,380	23,699
4	Shenzhen	China	22,940	22,570	22,510
5	Guangzhou	China	17,740	14,260	12,550
6	Busan	South Korea	17,040	16,170	14,194
7	Ningbo–Zhoushan	China	16,830	14,720	13,144
8	Qingdao	China	14,500	13,020	12,012
9	Dubai	United Arab Emirates	13,300	13,010	11,600
10	Tianjin	China	12,300	11,590	10,080
11	Rotterdam	Netherlands	11,866	11,880	11,140
12	Port Klang	Malaysia	10,001	9600	8870
13	Kaohsiung	Taiwan	9781	9640	9180
14	Hamburg	Germany	8863	9040	7910
15	Antwerp	Belgium	8635	8660	8470
16	Los Angeles	United States	8078	7940	7830
17	Dalian	China	8063	6400	5260
18	Tanjung Pelepas	Malaysia	7494	7500	6540
19	Xiamen	China	7202	6470	5820
20	Jakarta	Indonesia	6217	5620	4720

Source Wiki_ busiest container ports

in land scarce cities such as Singapore and Hong Kong. In these two cities, containers are stacked beyond their usual 3 or 5 high configurations, respectively. In Singapore, laden containers can be stacked up to 9 high and as such there is clearly a demand for OR techniques to be used in container extraction for loading or rapid container reassignment within the yard. Such techniques involve the storage policy of the containers, the movement equipment used, the location of storage, and the assignment of the berths (Bish et al. 2001).

This network effect goes beyond the individual countries and onto the region. We take a look at the situation of the container ports in the Pearl River Delta (PRD) region in China. This region has enjoyed a considerably advantageous position and experienced rapid development in terms of port container throughput growth, supported by large hinterland growth in factory production (Zhang et al. 2005). A similar situation exists in Southeast Asia as shown in Fig. 1.1.

As for the favorite OR techniques to use, there are plenty. In the literature, various OR techniques have been applied for a facet of container operations problems. The more common techniques include Mixed Integer Linear Programming (MILP), column generation, Genetic Algorithm (GA), Tabu search, simulation, and so on.

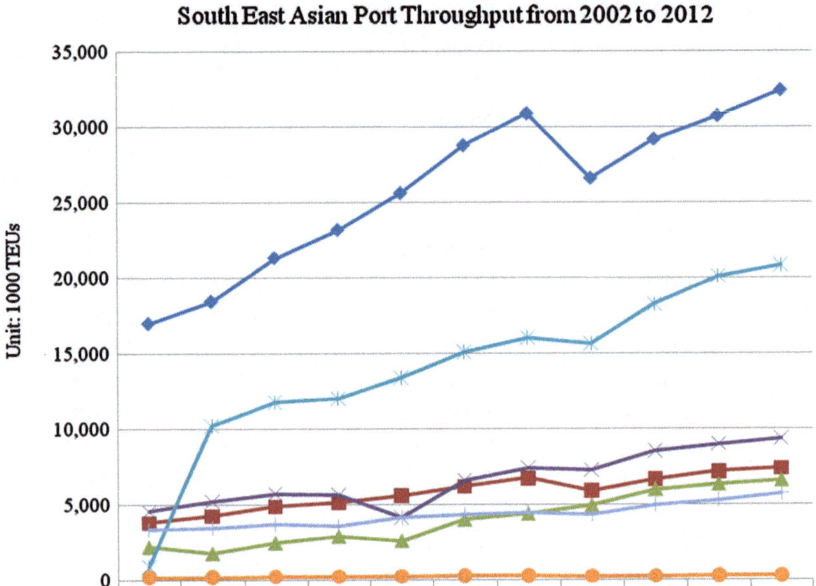

	2002	2003	2004	2005	2006	2007	2008	2009	2010	2011	2012
Singapore	16,986	18,441	21,329	23,192	25,608	28,768	30,891	26,593	29,179	30,728	32,422
Thailand	3,799	4,233	4,847	5,115	5,574	6,200	6,726	5,898	6,649	7,171	7,372
Vietnam	2,196	1,772	2,467	2,905	2,605	4,009	4,394	4,937	5,984	6,335	6,589
Indonesian	4,540	5,177	5,716	5,653	4,117	6,582	7,405	7,244	8,482	8,966	9,325
Malaysia	875	10,210	11,776	12,027	13,419	15,093	16,025	15,671	18,268	20,140	20,867
Cambodia	167	181	214	211	221	253	259	208	224	237	246
The Philippines	3,325	3,468	3,701	3,594	4,157	4,339	4,471	4,307	4,947	5,264	5,720

Fig. 1.1 South East Asian ports throughput and development over the last decade. *Source* Maritime Insight (2014)

References

Bish, E.K., Leong, T.Y., Li, C.L., Ng, J.W.C. and D. Simchi-Levi (2001). Analysis of new vehicle scheduling and location problem. Naval Research Logistics. 48: 363–385.

Maritime Insight (2014). Container Throughput and Economic Development, 2(3).

Stahlbock, R., Stefan Voß (2007). Operations research at container terminals: a literature update. OR Spectrum 30(1): 1–52.

Steenken, D., Stefan Voß and R. Stahlbock (2004). Container terminal operation and operations research—a classification and literature review. OR Spectrum 26(1): 3–49.

Wiki_ busiest container ports, List of world's busiest container ports, http://en.wikipedia.org/wiki/List_of_world's_busiest_container_ports. Accessed Feb. 14, 2015.

Wiki_Container terminal, http://en.wikipedia.org/wiki/Container_terminal. Accessed Feb. 14, 2015.

Zhang G., Zhang, N. & Q. Wang (2005). Container ports development and regional economic growth: an empirical research on the Pearl River Delta region of China, Proc. of the Eastern Asia Society for Transportation Studies, 5:2136–2150.

Chapter 2
Maritime Terminal Operational Problems

This Chapter provides an introduction to maritime terminal configuration and operational problems. The structure of typical container terminals and associated decision problems is discussed. A case study of one of the largest container terminals, PSA Singapore Terminals, is given. A literature review on selected key problems in container terminals such as berth allocation, quay crane scheduling, transportation, yard crane scheduling, and the integration of decision problems is conducted.

2.1 Container, Container Vessels, and Configuration of a Container Terminal

Malcolm McLean was the first freight forwarder who used specially designed containers on its vessel in 1956 (Frank 2009). Standard sized containers can be efficiently stacked and be moved seamlessly between ships, trucks, and trains. Vessels, trains, trucks, and cranes can be designed to a single size specification. The use of standard sized containers for cargo shipment has been proved to be able to substantially reduce the handling time of vessels and reduce the amount of labor needed. Containers, container vessels have gained popularity rapidly in the past several decades.

The basic container unit is of size 20′ × 8′ × 8′6″ (length × width × height), also referred to as a TEU (20-ft equivalent unit). Cargo volume and vessel capacity are usually measured in TEUs. The 20 and 40-ft lengths are the two most important and most commonly used sizes. The 20-ft container, referred to as a TEU, has become the industry standard reference. The 40-ft length container, also known as the 40-ft Equivalent Unit (FEU), has become the most frequently used container size today.

To enjoy economics of scale, the size of a container vessel has been increasing in the past few decades. In 2006, the world's largest containership, Emma Maersk, has a capacity of 11,000 TEUs (PSA International 2006). Today, the largest capacity of a container vessel has reached 19,224 TEUs (Stromberg 2015).

W. Li et al., *Planning and Scheduling for Maritime Container Yards*,
DOI 10.1007/978-3-319-17025-1_2

To meet the demand for an increasing number of container shipments, existing seaport container terminals were expanded, new container terminals, especially large ones, were constructed in recent years. Nowadays, the top container terminals in the world include Shanghai, Singapore, Hong Kong, and Shenzhen (Salisbury 2013).

A container terminal (CT) is a very important node in container transport, which is used mainly to serve container vessels. Vessels are loaded and unloaded in the CT and containers are temporarily stored before moving to their next mode of transport—rail, road, or sea. A CT is depicted in Fig. 2.1.

The layout of a typical container terminal is shown in Fig. 2.2.

Vessels mooring at the berths wait for quay cranes (QCs) to upload/discharge their containers. The PMs (prime mover) shuttle between the QCs and yard cranes (YCs) to move these boxes from the berth to the container yard area, and vice versa. Upon arrival at the blocks, the PMs queue in front of the YCs until they are served. Container yard areas serve as buffers where containers are temporarily stored. PMs also shuttle between YCs and gates to move boxes to/from YC to/from hinterland through gates. Figure 2.3 shows the container vessels mooring at a CT.

Three types of containers, export, import, and transhipment may coexist in a container terminal. The flows of the three types of containers are shown in Fig. 2.4.

Export containers are delivered from the hinterland by truck or train to the terminal gates. The containers are then checked and the transportation documents are processed at the gates before moving to the stack yards by the PMs. The YCs unload the containers from the PMs and put them into temporary storage in the yard stacks. When the vessel into which they are to be loaded arrives at the terminal,

Fig. 2.1 Corner of a CT (*Source* PSA Singapore Terminals)

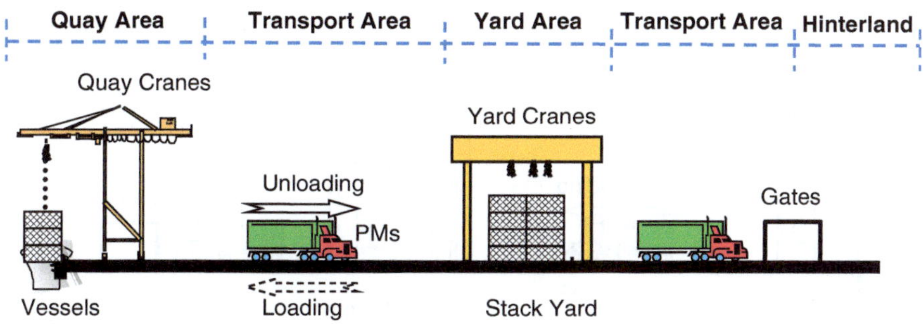

Fig. 2.2 Typical container terminal operations

Fig. 2.3 Vessels mooring at a CT (*Source* PSA Singapore Terminals)

these export containers are retrieved from the stack yards by the YCs and carried by the PMs to the berth. The PMs send the containers to the appropriate QCs which lift the containers and load them onto the mother vessel. These export containers are then delivered to their destination ports by the vessels.

Import containers arrive by vessel and leave the terminal via truck or train to the hinterland. An import container is first unloaded from a vessel by a QC which puts it onto a PM. The PM then moves the container to a storage yard where a YC unloads it from the PM and places it in the stack yard for temporary storage. Later, these import containers are retrieved from the stack yard by the YCs and carried by the PMs through the gate to its hinterland destination.

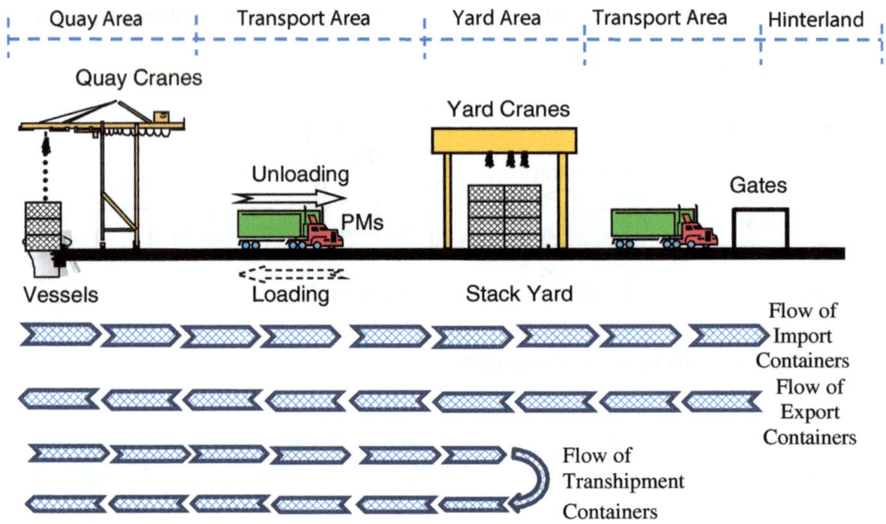

Fig. 2.4 Container flows in a CT

A transhipment container tranships from one (origin) vessel to another (destination) vessel with the help of a CT. Upon arrival of the origin vessel, a transhipment container is unloaded by a QC, moved from the quay to the yard by PMs, and then temporarily stored in a stack yard by a YC. Later when the destination vessel arrives at the terminal, the container is retrieved by a YC, moved from the yard to the quay and then loaded by a QC into the destination vessel. The destination vessel then unmoors from the berth and heads for its next port. A CT thus serves as a buffer (in space and time) between two container vessels.

A CT typically consists of several areas: quay area, transport area, yard area, and hinterland as shown in Fig. 2.2. The hinterland refers to the origin of export containers and destination of import containers. External trucks (hauliers) or trains deliver containers from/to inland origins/destinations. Figure 2.5 is a schematic diagram of a CT. Vessels, QCs, and YCs are labeled with numbers for easy identification. The small black rectangles represent the PMs.

In Fig. 2.6, we can find the quay area, transport area, and yard area of a CT.

We describe the quay area, transport area, yard area in more detail as follows.

2.1.1 Quay Area

The quay area provides the berths for vessels to moor. The quay cranes perform the loading and unloading operations of containers at a quay area. After the arrival of a vessel, a berth is assigned to the vessel. The QCs load/unload the containers which are delivered by the PMs from/to a yard area. A QC handles the containers in a

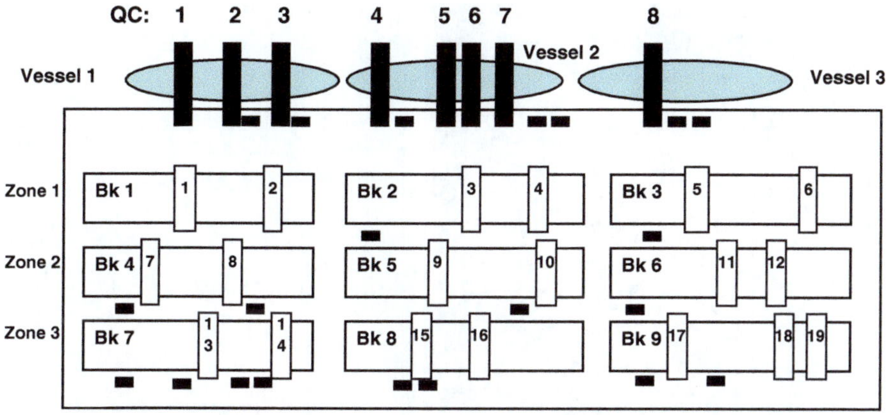

Fig. 2.5 Schematic diagram of a CT

Fig. 2.6 Quay area, transport area, and yard area of a CT (*Source* PSA Singapore Terminals)

different section along the length of the vessel known as a hatch. Typically, three to four QCs work on a vessel at a time. Figure 2.7a shows some QCs operating in a CT. In Fig. 2.7b, a QC is unloading a vessel and putting a container onto a PM.

There are two types of QCs: single trolley and dual trolley. A key indicator for the productivity of a quay area is the number of QC loading/unloading operations (called moves) per hour. QCs can operate 50–60 moves/h technically (Dirk et al. 2005). The actual performance of the QCs is, however, much lower: around 30 moves/h. The

Fig. 2.7 **a** QCs operating in a CT; **b** QC loading a PM (*Source* PSA Singapore Terminals)

average number of lifts achieved at a terminal per QC working hour is known as the GCR (gross crane rate). The GCR is a very important performance measure of a CT. Careful synchronization of the operations of PMs and YCs is the key to improving QC performance. The flow of containers should be as smooth as possible and the idle time of QCs waiting for yard trucks (YTs) should be minimized. This bears some similarity to the concept of JIT (Just-In-Time) where the flow of jobs should be smooth and leveled. We note in Chap. 3 that a CT is often bottlenecked by slow YC movements.

2.1.2 Transport Area

In transport area, containers are moved back and forth by the transport equipment. There are two transport areas: quayside transport area which is between the quay area and yard area, and the landside transport area which is between the yard area and the gates. At a CT, the transportation provided by PMs between QCs and stack yards or between stack yards and gates are usually referred to as horizontal transportation. The loading/unloading movements of the QCs and the YCs are referred to as stacking transport.

The most common PMs are the YTs (Fig. 2.8) and straddle carriers. The YTs are unable to lift the containers. Thus, the QCs and YCs are required to load/unload

Fig. 2.8 Yard truck pulling a container

containers from/to the trucks. On the contrary, the straddle carriers can lift the containers themselves, which can increase the productivity of the QCs. However, the straddle carriers require higher purchasing, maintenance, and operational costs (Frank 2009). The other types of equipment, such as the Automated Guided Vehicles (AGVs), are also often used. AGVs can reduce the labor cost of a terminal because they do not need drivers but their transport productivity and average speed are relatively low (Frank 2009). Nowadays, the AGVs are usually not used between gates and stack yards. Terminals may use any of the PMs mentioned above. In this book, we focus only on the YTs. Thus, the PMs in this book refer to the YTs.

2.1.3 Yard Area

A CT acts as a buffer between the various trade routes. Vessels from different trade routes rely on the buffer to maintain their schedules which have substantial economic impact for the global economy. The physical entity lying at the heart of this buffer is the yard area, which is used for intermediate storage of the containers. The export, import, or transhipment containers are temporarily stored in yard area before they are sent to their final destinations. The yard area is the buffer to link the different types of transportation vehicles or the same type of vehicles but arrive at different time periods. In some CTs such as PSA Singapore, the majority of yard activity comes from the transhipment operations. Figure 2.9 shows a yard area at PSA Singapore.

A yard area is typically partitioned as contiguous rectangular blocks (see Figs. 2.5 and 2.10) where a grid has been painted on the pavement indicating the positions where containers should be placed. In Fig. 2.5, we illustrate 9 yard blocks. Bidirectional traffic lanes for the PMs occupy the space between the blocks.

Fig. 2.9 Yard area at PSA Singapore (*Source* PSA Singapore Terminals)

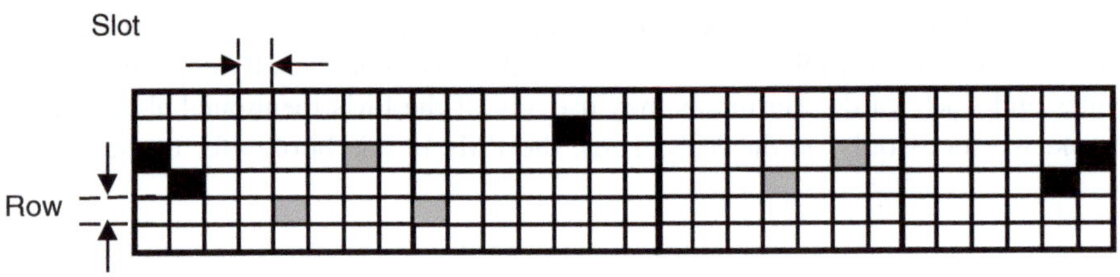

Fig. 2.10 Bird's-eye view of a typical yard block

A typical block is six stacks (6 × 8.5 feet) wide and thirty stacks (30 × 20 feet) long. Blocks are divided along their length into 20 ft sections called slots. Each slot has several rows. The containers are stored alongside in each row and are stacked on top of each other. A typical block is six rows (6 × 8.5 feet) deep and forty slots (40 × 20 feet) long.

The YCs perform the stacking and retrieval operations in this area. The YCs transfer the cargo into and out of the storage blocks. The YCs straddle above the containers in each block and traverses parallel to the length of a block. A yard zone is defined as a group of adjacent blocks that are aligned lengthwise and that together they form a single lane for yard crane movement. In Fig. 2.5, we illustrate three yard zones (e.g., block 1–3 are in zone 1; block 4–6 are in zone 2, etc.). A YC can access all storage locations in a yard zone without changing lanes. The YCs move easily within yard zones but take a long time to move between different zones. The movement of the YCs within (between) the zones is called a linear gantry (cross-gantry) movement.

Fig. 2.11 OBC moving a container in a yard area (*Source* PSA Singapore Terminals)

A yard area typically has space for many stacks of containers, with up to 6 containers per stack.

The YCs can be rail-mounted gantry cranes (RMGCs), rubber tired gantry cranes (RTGCs), or overhead bridge cranes (OBCs). Figure 2.11 shows an OBC moving a container inside a yard area. The RTGCs are more flexible in operation. To avoid operations interruptions, the neighboring YCs are kept apart from each other a safety distance, which is a multiple of a slot. The YCs operate approximately 20 moves/h technically (Dirk et al. 2005). The tightest constraint on the movement of the YCs and PMs is the slow gantry movement of the YCs and the extremely slow cross-gantry movement of the YCs. In particular, if a YC wishes to move from its current zone to another yard zone, the minimum amount of time required is around 15 min. In addition, transhipments are the most crucial operations as they involve the vessels directly. A delay in transhipment operations (especially during a vessel loading) carries a very high immediate cost.

2.1.4 Practical Example of CT: PSA Singapore Terminals

As the world's largest transhipment hub and the second largest container terminal (in terms of volume of TEUs handled) in the world, PSA Singapore Terminals

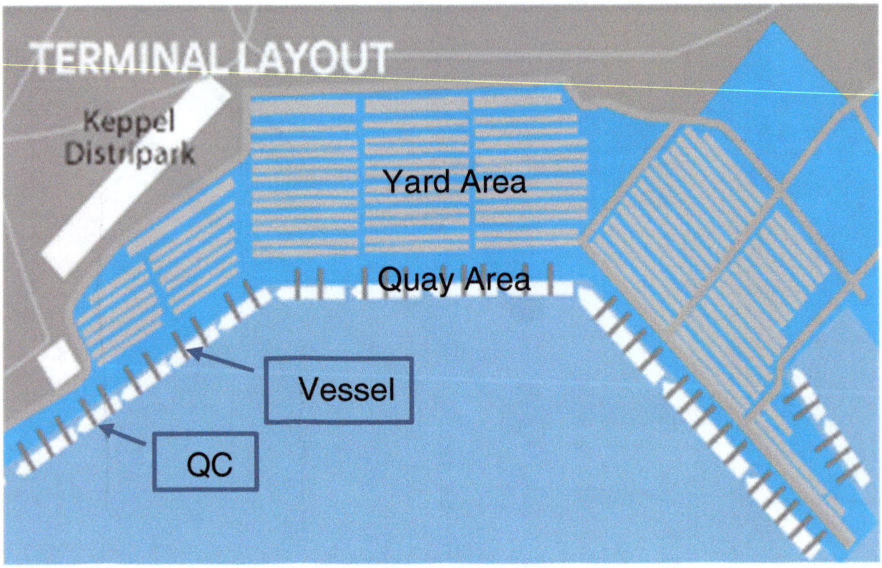

Fig. 2.12 Layout of Keppel terminal (*Source* Factsheet 2014)

handled 32.24 million TEUs in 2013 (Factsheet 2014) and 33.55 million TEUs in 2014 (PSA International 2015). It connects the cargo markets of the West (Europe, Africa, Middle East, and South Asia) with that of East Asia. PSA Singapore Terminals moves a seventh of the world's shipping containers (Credit Opinion 2014).

In Singapore, PSA Singapore Terminals operates seven container terminals (Tanjong Pagar, Keppel, Brani, Pasir Panjang Terminals 1, 2, 3, and 5) offering connections to 600 ports globally. There are a total of 57 container berths over a quay length of 17,350 m (Factsheet 2014). The layout of Keppel Terminal is depicted in Fig. 2.12. There are 14 container berths in its 3200 m quay length. A total of 40 QCs are working in this 105 hectare terminal (Factsheet 2014).

In Fig. 2.12, we can see a large yard area (with many long brown bars representing yard blocks). The quay area is set up along the seaside with white boxes representing the vessels and the short bars representing the QCs.

2.2 Decision Problems

With the ever increasing international container traffic, CT operations are becoming more and more important, busy, and complex. Competition among the CTs is also increasing: The CTs compete with each other in providing better customer service to attract more container shipments. The CT operators such as PSA Singapore face constant pressure to reduce the turnaround time and increase flow efficiency. Efficient computational tools which can help increase container throughput are key. Large CTs have reached a degree of complexity that further improvements require scientific methods (Steenken et al. 2005). Many publications on CT operations have appeared in the past several decades. These publications provide the different tools for different types of decision problems in the CTs.

The decision problems in a CT are categorized into several different categories by different researchers. Böse (2011) introduced a three level model for CT planning:

Level 1 (Planning of terminal infrastructure): deals with decisions regarding preparation of the terminal (area, quay wall length, etc.) and terminal connection to the external networks.
Level 2 (Planning of terminal suprastructure): deals with decisions regarding what kinds of resources and in what quantity should be used in a CT. For example, CT layout design.
Level 3 (planning of terminal operation): deals with short-, medium- and long-term planning of the CT's operations. Level 3 can be further broken down into short-term, medium-term, and long-term planning.

There are interdependencies among the three levels. A lower level provides support and restriction for the higher levels. A higher level generates the requirements to the lower levels. Basically, level 1 is the duty of the local communities and authorities, which is basically out of the scope of academic research (Böse 2011). Levels 2 and 3 are areas of abundant research. Böse (2011) has compiled the research work on level 2: planning of terminal suprastructure.

Günther et al. (2005) categorized the decision problems in a CT into three categories: design, planning, and real-time control problems. The design problems involve the types and the quantity of resources in a CT such as the type of handling equipment in the yard, the number of berths, QCs, YCs, YTs, storage slots, the yard layout, and the human operators, etc. Planning problems deal with how to schedule operations in a CT in advance to better utilize resources to maximize the efficiency of the operations. Real-time control deals with the assignment of resources to tasks in real time. The detailed operations schedules can only be developed in a very short time (several minutes) for some resources, such as the PMs, and the YCs. Real-time control necessitates very fast algorithms without sacrificing too much on the optimality.

The design problems found in Günther et al. (2005) corresponds to level 2 (planning of terminal suprastructure) in Böse (2011). The planning and real-time control problems roughly belongs to level 3 (planning of terminal operations) in Böse (2011).

In this book, we apply the categories proposed by Günther et al. (2005) because level 1 in Böse (2011) is basically out of the scope of our research. We focus on the planning and real-time control problems in this book.

2.3 Literature Review

In this section, we provide a literature review on selected key decision problems at different areas. Meersmans (2002), Vis and de Koster (2003), Steenken et al. (2004), Vacca et al. (2007), Stahlbock and Voß (2008), and Gharehgozli et al. (2014a) provide excellent reviews of the extant literature on container terminal operations.

2.3.1 Quay Area

The quay area decision problems are critical and have a major impact on a CT's operational performance. In the quay area, the corresponding decision problems include berth allocation, quay crane assignment, quay crane scheduling, and stowage planning. Stowage planning deals with how to assign the containers to empty slots in a vessel. We will not review the work on stowage planning.

2.3.1.1 Berth Allocation

A specific anchoring berth has to be allocated to a vessel by the CT planners before the vessel's arrival. The Berth Allocation Problem (BAP) deals with how to optimally allocate such vessels to the berths or quay locations. The berthing (handling) time and the berthing position along the quay length of a vessel have to be determined. The input data to be considered include the technical specifications of the vessels (e.g., vessel length, arrival time), QCs (e.g., length of crane jib), data on the berth type (e.g., berth layout, number, length, etc.), projected vessel handling time, mooring time windows, priorities of the vessels, dedicated berth areas, etc. The objectives of BAP include maximizing the productivity of the vessel handling, maximizing customer service levels, minimizing the total service time of vessels, and minimizing the costs.

Lim (1998) formulated the berth planning problem as a mathematical model and proved the problem's NP-completeness. The problem was then transformed into a two-dimensional packing problem and a heuristic was proposed to effectively solve it. Nishimura et al. (2001) treated the berth allocation problem as a public berth system and applied genetic algorithms to obtain computationally efficient solutions. Guan and Cheung (2004) consider BAP with an objective of minimizing the total weighted flow time. Composite heuristics are proposed for large size problems.

Most BAP treat each arriving vessel equally. However, in actual operations, some vessels desire a high priority service and need to be allocated to the berths earlier (Imai et al. 2003). The vessels with a large container handling volume generally are also given a higher priority over the smaller and less frequent vessels. In general, ship priority depends on the total throughput per shipping line (Imai et al. 2003). Thus, some researchers have incorporated service priority into the

BAP. Imai et al. (2003) modified the BAP formulation to deal with the calling vessels with inhomogeneous service priorities. Instead of the assigning weights to represent the vessel priorities, Golias et al. (2009) developed a multi-objective BAP. Different groups of vessels use different objective functions in their models. Vessel service is differentiated based on certain priority agreements. For other recent research on BAP, the interested reader is referred to Hendriks et al. (2010), Buhrkal et al. (2011), Du et al. (2011), Hendriks et al. (2012), and Xu et al. (2012).

2.3.1.2 Quay Crane Assignment

After assigning the berths to the vessels, we now need to allocate the QCs to the vessels for loading/unloading operations which are done through the Quay Crane Assignment (QCA) problem. The QCA is sometimes referred to as a crane split (Stahlbock and Voß 2008). It deals with the allocation of the QCs to the vessels and the vessels' bays. Given the vessels to be served and available QCs to be used, the QCA has to ensure that the overall loading/unloading operations of vessels can be completed as quickly as possible. The QCA is closely related to the BAP: The solution to the BAP serves as the input to the QCA, while the solution from the QCA will in turn affect the BAP.

Peterkofsky and Daganzo (1990) pointed out the importance of crane operational efficiency and developed a branch and bound solution to speed up a cargo ship's loading and unloading operations. Park and Kim (2003) formulated an integer programming model for the quay crane allocation problem with various practical constraints and designed a two-phase solution procedure for the mathematical model. Chang et al. (2010) applied objective programming to a dynamic allocation model for berth allocation and quay crane assignments based on a rolling-horizon approach. Due to the computational complexity of the problem, a hybrid parallel genetic algorithm was designed to solve the problem and its effectiveness was evaluated via simulation. Han et al. (2010) simultaneously addressed the berth and quay crane scheduling problems and took vessel arrival times and container handling uncertainty into consideration. A mixed integer programming model was proposed and a simulation-based genetic algorithm procedure was employed to provide the proactive berth and quay crane schedules. Giallombardo et al. (2010) combined the berth allocation and the quay crane assignment problems and developed a mixed integer quadratic program which was subsequently reduced to a mixed integer linear program. A heuristic which combined tabu search and mathematical programming techniques was developed to solve the problem.

2.3.1.3 Quay Crane Scheduling

The QCA allocates the available QCs to the vessel bays. The next decision is to decide the sequence of the loading/unloading tasks performed by these QCs, which is done by Quay Crane Scheduling (QCS). The tasks to the QCs are the containers on the vessel to be unloaded and the containers on the PMs delivered from the yard area to be loaded. A QCS model typically consider the following constraints: unloading tasks must precede any loading task, interference between neighboring QCs, a QC can perform at most one task at a time, and the QC travel speed. The objective is usually to minimize the makespan of all tasks.

Kim, Kang, and Ryu (2004) considered the load sequencing of outbound containers. They decompose the problem into two subproblems: pickup schedule and load sequence. A beam search algorithm was used to solve the model.

In the QCS model developed by Lim et al. (2004), spatial and separation constraints were considered. In their QCS model, QCs cannot cross over; there is a minimum distance between the cranes and jobs cannot be done simultaneously. Dynamic programming algorithms, a probabilistic tabu search and a squeaky wheel optimization heuristic were proposed to solve the model. Lim et al. (2007) further developed a scheduling procedure which assigns the containers to non-crossing quay cranes to minimize the maximum completion time of all operations. The problem was decomposed into start-time allocation and job-to-crane allocation. A dynamic programming, backtracking exact algorithms, and a simulated annealing algorithm were employed to solve the proposed problem.

The QCS problem has also been studied by Kim and Park (2004) for quay cranes operating with non-crossing constraints, safety distance (at least one ship-bay), precedence constraints among the tasks in the same bay, and a consideration of container groups. A mixed integer programming model, a branch-and-bound algorithm, and a greedy randomized adaptive search procedure (GRASP) were proposed in order to minimize the total completion time. Moccia et al. (2006) strengthened the formulation of Kim and Park (2004) and developed a branch-and-cut algorithm which outperformed Kim and Park's branch and bound approach. Sammarra et al. (2007) used tabu search to study the same problem as defined by Kim and Park (2004), and evaluated the performance of the tabu search approach against the branch-and-cut algorithm presented by Moccia et al. (2006), and the GRASP approach proposed by Kim and Park (2004). Bierwirth and Meisel (2009) resolved the problem of Kim and Park (2004) in a limited solution space by using unidirectional schedules (i.e., the QCs are only allowed to move in one direction) and compared their proposed branch-and-bound approach with the results of Kim and Park (2004), Moccia et al. (2006), and Sammarra et al. (2007). Kaveshgar et al. (2012) used a genetic algorithm to cope with the drawback of Bierwirth and Meisel (2009) model where the QCs can only transfer in one direction.

The problem of identical quay crane scheduling has been discussed by Ng and Mak (2006) who introduced a decomposing scheduling heuristic which divides the vessel into nonoverlapping zones, and a dynamic programming procedure was used to obtain the optimal partitioning. Zhu and Lim (2006) formulated the QCS problem as an integer programming model to minimize the overall makespan where the non-crossing constraint was considered, and all jobs in a particular bay need to be finished before the QC could move to another bay. The online version of the work of Zhu and Lim (2006) was followed by Zhang et al. (2008) and two heuristics, namely, (1) one-by-one or online list (i.e., without any available information about the remaining operation for consideration) and (2) release dates or online time (with regard to available information for release date) were proposed.

Lee et al. (2008b) focused on the QCS problem for a single vessel using handling priority for each ship bay and non-crossing operational restrictions. A mixed integer programming (MIP) model and a genetic algorithm (GA) were presented to minimize the total weighted completion time of every ship bay. In a similar study by Lee et al. (2008c), the QCS problem with non-crossing constraints was addressed for a container ship.

Tavakkoli-Moghaddam et al. (2009) tried to solve the quay crane assignment (QCA) and the quay crane scheduling (QCS) problems simultaneously. An MIP model and GA were used to analyze real-sized problems and to optimize the total completion time of the vessels and the QCs. In comparison with the results for the MIP model provided by the LINGO optimization software package, it was found that the proposed GA was better in terms of the solution times.

A new model for the QCS problem has been developed by Meisel (2011) with respect to the time window constraint for the quay cranes. A mathematical formulation, a lower bound and the unidirectional search heuristic (which is an extension from Bierwirth and Meisel 2009) were used to deal with the minimization of the total vessel handling time.

Legato et al. (2012) presented a QCS model which includes different processing times for quay cranes for the container group as well as nonidentical QCs, non-crossing constraints, time windows, and individual unidirectional schedules. They formulated a MILP, a Lagrangian relaxation approach for finding a lower bound, and an enhanced branch-and-bound algorithm based on the work of Bierwirth and Meisel (2009) to solve the scheduling problem.

Lu et al. (2012) studied the QCS problem by considering multiple QCs which are able to process different operations (but not simultaneously) at a single bay (shared bay). A polynomial time complexity heuristic approach was developed to find the optimal assignment and sequencing of containers to contiguous QCs.

2.3.2 Transport Area

The transport area serves as the link between the quay area and yard area, and between the yard area and the hinterland. The decisions on the transport area aim to reduce the transport times and ensure QC and YC productivity. High-speed QCs or YCs do not necessarily mean high productivity in a CT. Increasing the number or the speed of PMs may not always improve the performance of a CT. Careful synchronization among the transport, quay, and the yard areas is crucial to improving a CT's performance. The decision problems in the transport area include the allocation of the PMs to the QCs, the allocation and sequencing of jobs to the PMs, and the routing of the PMs. The objectives in optimizing transport operations can be to minimize the transport times, minimize the QC idle times, or minimize the vehicle fleet size.

Nishimura et al. (2005) focused on PM control and considered PM routing. However, the study assumed a relatively static environment where each PM travels along a cyclical path with limited opportunities for rerouting.

There are many papers on AGV dispatching that are in the same vein as PM routing and scheduling. van der Meer (2000) provided a comprehensive introduction to the AGV dispatching methods. The research on multiple-load AGV systems commenced with Bilge and Tanchoco (1997), with contributions from a handful of other authors, including Sinriech and Kotlarski (2002). Levitin and Abezgaouz (2003) considered multiple-load AGVs with LIFO loading constraints. However, their study only considered a single vehicle operating in isolation. van der Meer (2000), Vis and Harika (2004), and Yang et al. (2004) considered a system of automated lifting vehicles (ALVs) for the internal transportation of cargo in container terminals. Evers and Koppers (1996), Cheng (2001), Ioannou et al. (2001), Kim and Bae (2004), among others, studied the AGV system in a CT where a single-load AGV capacity was assumed.

For multi-load AGV systems in a CT, Chan (2001) used simulation to compare two alternative dispatching strategies for the case where there are n single-level, multi-load AGVs and m containers waiting to be moved. However, the assumption that all m jobs are ready at time 0 does not reflect the true nature of CT operations. Grunow et al. (2004) studied the dispatching of single-level, dual-load AGVs in an automated CT similar to ECT Rotterdam and CTA Hamburg. The AGVs can accommodate one 40′/45′ container or up to two 20′ containers. In these studies, the details of AGV routing and traffic control were ignored and only dispatching was studied. After carefully establishing the ideas of full and partial availability of the AGVs, the authors developed a novel heuristic dispatching algorithm. This algorithm was initiated whenever a new transport order appears within the look-ahead time window. For each partially available AGV, the algorithm generates up to three possible tours for the AGV, each one corresponding to a different ordering of the pick-up/drop-off operations of the new container with respect to the drop-off of the container already on the AGV. One noteworthy finding in Grunow et al. (2004) is that the benefit of dual loading on the AGVs diminishes as the CT becomes larger.

Nguyen and Kim (2009) discussed the dispatching of ALVs. An MIP model was developed to assign delivery tasks to the ALVs optimally. Petering (2010) simulated a real-time dual-load yard truck control in a transhipment terminal.

2.3.3 Yard Area

Yard area operations are often a potential bottleneck in a CT (Li et al. 2009). The performance of a CT depends heavily on the decisions made for yard area operations. In the yard area, the corresponding decision problems include assigning the yard blocks for the calling vessels, determining the storage locations for individual containers, yard remarshalling, and Yard Crane Scheduling (YCS). The YCS has two subproblems: YCS-I which deals with assigning the YCs to several yard blocks and YCS-II which deals with scheduling individual YC storage/retrieval operations within a block. In this book, we focus on YCS-II which has some unique features that make it unlike the other problems often studied with respect to material handling systems in the literature. For example, two working YCs in the same lane must remain separated by at least 160 feet; the YCs interact with the PMs only and there is no buffer at the YC/PM interface. A PM must be present at a location in order for a YC to complete handling at that location. YCS-II is also affected by the QC operations: The QC work list determines the target times and locations of storage and retrieval container moves. While the literature contains many articles on CT operations, only a few studies focus on YCS-II. We will provide more literature review on YCS-II in Chaps. 3–6.

For YCS-I, Linn et al. (2003) introduced an MIP model to optimally deploy the YCs within a storage yard including cross-gantry moves. The aim of this YC deployment procedure was to move the idle YCs to the other blocks at the beginning of the next planning horizon to even the YC workloads. Linn and Zhang (2003) studied the problem of finding an optimal number of YCs in each yard block for every planning period with cross-gantry moves and developed an MIP model and a least cost heuristic to minimize the YC workload overflow.

Legato et al. (2009) investigated the deployment of multiple YCs among the yard blocks with particular respect to the linear-and cross-gantry movements. An integer programming model and a discrete event simulation were proposed in order to find the block pair for each YC, and to rank and select five proposed cross gantry deploying policies. The numerical results showed that the "crane transfer to the closet yard block" policy outperformed the other proposed configurations and rules.

Petering et al. (2009) studied the impact of a real-time YC dispatching system in respect of the long-term performance of a CT with the aim of obtaining a high quay crane usage rate in a transhipment terminal. Twelve rule-based yard crane dispatching algorithms and look-ahead yard crane dispatching systems were used in a discrete event simulation model to evaluate their performance.

The problems of linear- and cross-gantry interference have also been considered by He et al. (2010) in order to optimize the productivity of yard crane operations. Their proposed model minimized the total number of delayed jobs and the total time of YC inter-block moves by proposing a hybrid algorithm based on seven heuristics and a parallel genetic algorithm. Yan et al. (2011) used a knowledge-based approach for the YCS problem and improved the results compared with the approach previously adopted by He et al. (2010). Chang et al. (2011) presented a model for the YCS problem which assumed up to two YCs per block, cross gantry moves, and container groups to minimize the total job delay at the blocks. A dynamic rolling-horizon decision procedure and an integer programming model were developed to solve this problem. An integrated heuristic algorithm and simulation model were proposed, and a GA, which used the initial results from the heuristic algorithm, was introduced to cope with the scheduling problem. The computational results and simulation indicated that proposed procedure outperformed the static rolling-horizon technique presented by He et al. (2010).

Guo et al. (2011) noted that the problem of yard crane scheduling is NP-hard and hence adopted a three-level hierarchical procedure for yard crane operation management, namely, (1) the distribution of the YCs to different rows, (2) dispatching the YCs in a row, and (3) YC dispatching in individual zones. Within this approach, their study focused on the single yard crane dispatching at Level 3. An evaluation of the two new heuristic algorithms that were used to minimize the average vehicle delay at the yard showed to be promising. In a follow-up study, Guo and Huang (2012) focused on Level 2 of the hierarchical procedure in Guo et al. (2011), and proposed a time partition algorithm and a space partition algorithm to arrange the YC workload in a single row of yard blocks. It was found that the use of a time partitioning algorithm outperforms that of a space partitioning algorithm.

The real-time multiple YCS problem, and the associated consideration of loading operations and linear gantry movements has also been studied by Chen and Langevin (2011) in order to minimize the completion time of the handling of all the outbound containers. The model considered a minimum safety distance of five yard-bays and inter-crane interference between two adjacent yard cranes, also a maximum of two yard cranes in each block at any time. An MIP model was constructed, and GA and tabu search were utilized to obtain near-optimal results.

Sharif and Huynh (2012) investigated both the centralized and decentralized methods for scheduling multiple YCs. A comparison of the efficiency of these two methods in respect of five solution qualities (optimality, scalability, computational efficiency, adaptability, and fault tolerance) in truck operations showed that the centralized method outperformed the decentralized method by an average of 16.5 % but the decentralized approach demonstrated strong local adaptability to real-time truck arrivals.

For the YCS-II problem, Ng and Mak (2005a, b) addressed the problem of scheduling a single YC carrying out a given set of jobs with different ready times in order to minimize the total job waiting times. To achieve this, Ng and Mak (2005a) first developed a heuristic and a lower bounding algorithm to benchmark the performance of the heuristic. A branch-and-bound algorithm was then proposed in Ng and Mak (2005b). In addition, Ng (2005) presented an integer program, a dynamic programming-based heuristic, and a lower bound to solve the problem of scheduling multiple YCs. While the inter-crane interference was considered; there was no consideration of the safety distance requirement and no differentiation of the storage and retrieval jobs.

The single block twin-crane scheduling problem with both storage and retrieval jobs was also investigated by Vis and Carlo (2010). They constructed a mathematical model to minimize the makespan for both cranes. An algorithm to derive a lower bound for the makespan was then introduced and a simulated annealing-based heuristic was proposed to solve the problem. Stahlbock and Voß (2010) investigated the effectiveness of double-rail-mounted gantry cranes via simulation. They evaluated different online algorithms for the sequencing and scheduling of jobs for the automated double-rail-mounted gantry cranes at a storage block for instances derived from actual terminal operations.

Park et al. (2010) utilized the heuristic-based and local-search-based real-time scheduling methods for twin rail-mounted gantry cranes working in a block at an automated container terminal. They rolled planning horizon ahead whenever a crane finishes one job and therefore claimed the approaches are suitable for real-time scheduling. Specifically, they singled out the containers that require rehandling and treated them as independent jobs, which greatly facilitated the cooperation of the quay cranes and addressed the workload balancing and crane interference problems.

Recent container handing technologies, such as the double-rail-mounted gantry (DRMG) crane system, have been used in practice to improve the handling efficiency. The DRMG system consists of two rail-mounted gantry cranes of different height and width which allow them to cross each other. Cao et al. (2008) focused on providing an operational strategy for the system to load outbound containers. They proposed an integer programming model and then applied a greedy heuristic, a simulated annealing algorithm and a combined scheduling heuristic to tackle the problem. Dorndorf and Schneider (2010) scheduled triple crossover stacking cranes in an automated container storage block with asynchronous handover at the transfer areas at both block front ends. They treated the problem as an online optimization problem and constructed a new crane schedule for a certain planning horizon whenever a new job arrives or a job is completed.

Gharehgozli et al. (2014b) modeled the one-YC scheduling problem with both the storage and retrieval jobs in a single container block into a continuous time integer programming problem. A two-phase solution method, which utilized the intrinsic properties of the model, was adopted to optimally solve the problem. The first phase included a merging algorithm which tries to patch sub-tours into a complete crane tour taking consideration of extra travel time, while the second phase used a branch-and-bound algorithm to reach optimality.

We developed discrete and continuous time models for YCS-II in Chaps. 3, 4. Realistic operational constraints such as non-passing YCs, the safety distance between two YCs, and simultaneous storage/retrieval moves have been considered. High-quality solutions are obtained for actual problems with very short solution times.

2.3.4 Integration of Decision Problems

The three-level model introduced by Böse (2011) for CT planning are hierarchically interconnected: the lower level provides support and restrictions for the higher levels. The higher level generates requirements to the lower levels. This is also the case for the three category decision problems introduced by Günther et al. (2005). Even at the same level, the decision problems are also closely related. The solution to the BAP can serve as the input to the QCA, the solution to the QCA can in turn serve as an input to the QCS. However, the solution from the QCA and QCS will affect the BAP: the allocation of the QCs to the vessels and the detailed schedule of the QCs will affect the vessel handling time of the vessels, which is often assumed to be constant in BAP. The productivity of the QCs cannot be high if the efficiency of the PMs or YCs are low. A QC or YC will have to wait if the PMs arrive late. Similarly, the PMs have to wait if the QCs or YCs are busy. Thus, the operations of the quay, yard, and the transportation areas have to be carefully coordinated. The flow of containers should be as smooth as possible. Only through a systematic integration of all areas, a CT is able to achieve a high performance and fulfill its overall function as a key node in international transportation networks.

This interdependency nature calls for the integration of decision problems: at the hierarchical level (i.e., the decision problems of design, planning, and real-time control be integrated), and at the area level (i.e., the decision problems of quay, yard, and transportation areas be integrated). However, due to the extreme complexity of the decision problems at the different levels and areas, the model size often becomes intractable. There are not many research work considering the integration of such decision problems. Most research only consider the optimization of an individual problem in one area as listed in Sects. 2.3.1–2.3.3.

Recently, more researchers realized the importance of integrating the decision problems. Meersmans (2002) considered the integrated scheduling of automated stacker cranes (ASCs), AGVs and QCs at a container terminal. Some algorithms including branch and bound and beam search heuristics are tested. Bish (2003) considered the combined problems of QCs loading and unloading sequence, YT dispatching, and container storage locations at yards. The objective is to minimize the maximum time taken to serve a given set of vessels. A heuristic was proposed to solve the model. Chen (2007) presented integrated model to schedule various types of equipment at the same time. Both the loading and unloading operations are considered simultaneously. The problem is formulated as a job shop scheduling problem and solved with a tabu search algorithm. Lee et al. (2008a) considered the integrated scheduling of cranes and YTs. The problem is formulated as an MIP model and solved with GA to minimize the makespan of the cranes' tasks. Lau and Zhao (2008) addressed the integrated scheduling of QCs, AGVs and automated yard cranes (AYCs). They proposed a heuristic called GA plus maximum matching (GAPM) to obtain near-optimal solution. Cao et al. (2010) developed an integrated model for the yard truck and yard crane scheduling problems for loading operations. Solution methods based on Benders decomposition are proposed to solve the model. Yuan et al. (2011) proposed the integrated straddle carrier path planning and task allocation in yard operations including the QC and YT related jobs. A job grouping strategy was proposed to solve the problem.

Chen et al. (2012) studied BAP, QCA, and QCS in an integrated model and applied combinatorial Benders cuts algorithm to solve the model. Homayouni et al. (2013) formulated a model to optimize the coordinated scheduling of cranes and vehicles in CTs. The objectives are to minimize the total travel time of the trucks and delays in the tasks of the cranes.

Meisel and Bierwirth (2013) proposed a framework to integrate the BAP, QCA, and QCS models. They used three phases to solve the three subproblems. Problem scenarios of realistic size can be solved in their integrated framework.

Chen et al. (2013) considered the simultaneous scheduling of QCs, YTs, and YCs so that the makespan for serving a set of vessels is minimized. The problem is formulated as a constraint programming model. A three-stage algorithm is developed to combine the different subproblems. Tierney et al. (2014) developed an integer programming model for analyzing inter-terminal transportation (ITT) in sea ports. The movement of the containers between terminals (sea, rail, or otherwise) within a port was analyzed and modeled. A two-step solution procedure was proposed to solve the overall ITT problem. Homayouni et al. (2014) used GA to solve the integrated scheduling of the QCs, AGVs, and YCs. They showed that GA outperforms the simulated annealing algorithm for the problem.

Acknowledgments The authors would like to acknowledge the valuable suggestions and data provided by PSA Corporation Ltd., which have helped to improve this chapter.

References

Bierwirth, C., F. Meisel. (2009) A fast heuristic for quay crane scheduling with interference constraints. Journal of Scheduling. 12(4): 345–360.

Bilge, U. and Tanchoco, J. M. A. (1997) AGV systems with multi-load carriers: basic issues and potential benefits. Journal of Manufacturing Systems. 16:159–174.

Bish, E.K. (2003) A multiple-crane-constrained scheduling problem in a container terminal. European Journal of Operational Research. 144(1): 83–107.

Böse, J. (2011) Handbook of Terminal Planning, New York : Springer.

Buhrkal, K., S. Zuglian, S. Ropke, J. Larsen, R. Lusby. (2011) Models for the discrete berth allocation problem: A computational comparison. Transportation Research Part E: Logistics and Transportation Review. 47(4): 461–473.

Credit Opinion (2014) PSA Corporation Limited, Moody's Corporation, Moody's Investors Service, Inc., Oct. 19, 2014.

Cao, J. X., Lee, D-Hg, Chen, J. H., Shi, Q. X. (2010) The integrated yard truck and yard crane scheduling problem: Benders' decomposition-based methods, Transportation Research Part E: Logistics and Transportation Review. 46(3): 344–353.

Cao, Z., D. H. Lee, Q. Meng. (2008) Deployment strategies of double-rail-mounted gantry crane systems for loading outbound containers in container terminals. International Journal of Production Economics. 115(1): 221–228.

Chan, S. H. (2001) Dynamic AGV-container job deployment strategy. Master of Science, National University of Singapore.

Chang, D., Z. Jiang, W. Yan, J. He. (2010) Integrating berth allocation and quay crane assignments. Transportation Research Part E: Logistics and Transportation Review. 46(6): 975–990.

Chang, D., Z. Jiang, W. Yan, J. He. (2011). Developing a dynamic rolling-horizon decision strategy for yard crane scheduling. Advanced Engineering Informatics 25(3): 485–494.

Chen, L., Bostel, N., Dejax, P., Cai, J., & Xi, L. (2007) A tabu search algorithm for the integrated scheduling problem of container handling systems in a maritime terminal. European Journal of Operational Research. 181(1): 40–58.

Chen, L., A. Langevin. (2011) Multiple yard cranes scheduling for loading operations in a container terminal. Engineering Optimization 43(11): 1205–1221.

Chen, L.; Langevin, A. & Lu, Z. (2013) Integrated scheduling of crane handling and truck transportation in a maritime container terminal. European Journal of Operational Research, 225: 142–152.

Chen, J. H., Lee, D. H., Cao, J. X. (2012) A combinatorial benders' cuts algorithm for the quayside operation problem at container terminals. Transportation Research Part E: Logistics and Transportation Review. 48(1): 266–275.

Cheng, Y. L. (2001) Simulation study of a dynamic AGV-container job deployment scheme. Master of science, National University of Singapore.

Dirk Steenken, Stefan Voß, Robert (2005) Container terminal operation and operations research - a classification and literature review, In: Günther, Hans-Otto, Kim, Kap Hwan, Container Terminals and Automated Transport Systems.

Dorndorf, U., F. Schneider. (2010) Scheduling automated triple cross-over stacking cranes in a container yard. OR Spectrum. 32(3): 617–632.

Du, Y., Q. Chen, X. Quan, L. Long, R. Y. K. Fung. (2011) Berth allocation considering fuel consumption and vessel emissions. Transportation Research Part E: Logistics and Transportation Review. 47(6): 1021–1037.

Evers, J. J. M. and Koppers, S. A. J. (1996) Automated guided vehicle traffic control at a container terminal. Transportation Research Part A. 30(1):21–34.

Factsheet (2014) PSA Singapore Terminals, http://www.internationalpsa.com/factsheet/pdf/Singapore.pdf. accessed Feb. 14, 2015.

Frank, M. (2009) Seaside Operations Planning in Container Terminals, Heidelberg: Physica-Verlag.

Gharehgozli, A. H., Roy, D., & de Koster, M.B.M. (2014a) Sea Container Terminals: New Technologies, OR models, and Emerging Research Areas (No. ERS-2014–009-LIS). ERIM report series research in management Erasmus Research Institute of Management. Retrieved from http://hdl.handle.net/1765/51656.

Gharehgozli, A. H., Yu, Yugang, Koster, R. d., Udding, J. T. (2014b) An exact method for scheduling a yard crane. European Journal of Operational Research. 235(2): 431–447.

Giallombardo, G., L. Moccia, M. Salani, I. Vacca. (2010) Modeling and solving the tactical berth allocation problem. Transportation Research Part B: Methodological. 44(2): 232–245.

Grunow, M., Gunther, H.-O., and Lehmann, M. (2004) Dispatching multi-load AGVs in highly automated seaport container terminals. OR Spectrum. 26: 211–235.

Guan Y., and Cheung, R. K. (2004) The berth allocation problem: models and solution methods, OR Spectrum. 26: 75–92.

Günther, H.-O., Kim, K. H. (2005) Logistics control issues of container terminals and automated transportation systems, In: Günther, H.-O., Kim, K. H., Container Terminals and Automated Transport Systems.

Guo, X., Huang. S. Y. (2012) Dynamic space and time partitioning for yard crane workload management in container terminals. Transportation Science. 46(1): 134–148.

Guo, X., Huang, S. Y., Hsu, W. J., and Low, M. Y. H. (2011) Dynamic yard crane dispatching in container terminals with predicted vehicle arrival information. Advanced Engineering Informatics. 25(3): 472–484.

Han, X. L., Z. Q. Lu, L. F. Xi. (2010) A proactive approach for simultaneous berth and quay crane scheduling problem with stochastic arrival and handling time. European Journal of Operational Research. 207(3): 1327–1340.

He, J., D. Chang, W. Mi, W. Yan. (2010) A hybrid parallel genetic algorithm for yard crane scheduling. Transportation Research Part E: Logistics and Transportation Review. 46(1): 136–155.

Hendriks, M., M. Laumanns, E. Lefeber, J. T. Udding. (2010) Robust cyclic berth planning of container vessels. OR Spectrum. 32: 501–517.

Hendriks, M. P. M., D. Armbruster, M. Laumanns, E. Lefeber, J. T. Udding. (2012) Strategic allocation of cyclically calling vessels for multi-terminal container operators. Flexible Services and Manufacturing Journal. 24:248–273.

Homayouni, S. M., Tang, S. H., Motlagh, O. (2014) A genetic algorithm for optimization of integrated scheduling of cranes, vehicles, and storage platforms at automated container terminals, Journal of Computational and Applied Mathematics. 270: 545–556.

Homayouni, S. M., Tang, S. H. (2013) Multi Objective Optimization of Coordinated Scheduling of Cranes and Vehicles at Container Terminals, Mathematical Problems in Engineering, Article ID 746781, 9 pages, http://dx.doi.org/10.1155/2013/746781.

Imai A., Nishimura E., Papadimitriou S. (2003) Berth allocation with service priority. Transportation Research-B. 37(5): 437–457.

Ioannou, P. A., Jula, H., Liu, C. I., Vukadinovic, K., Pourmohammadi, H., and Dougherty, E. (2001) Advanced material handling: automated guided vehicles in agile ports. Final Report prepared for Task 1.2.6.1 for the Center for Commercial Deployment of Transportation Technologies.

Kaveshgar, N., Huynh, N., and Rahimian, S. K. (2012) An efficient genetic algorithm for solving the quay crane scheduling problem. Expert Systems with Applications. 39(18): 13108–13117.

Kim, K. H. and Bae, J. W. (2004) A look-ahead dispatching method for automated guided vehicles in automated port container terminals. Transportation Science. 38:224–234.

Kim, K. H., Kang, J. S., Ryu, K. R. (2004) A beam search algorithm for the load sequencing of outbound containers in port container terminals. OR Spectrum. 26(1): 93–116.

Kim, K. H. and Park, Y. M. (2004) A crane scheduling method for port container terminals, European Journal of Operational Research. 156: 752–768.

Lau, H. & Zhao, Y. (2008) Integrated scheduling of handling equipment at automated container terminals, International journal of production economics. 112(2): 665–682.

Lee, D.-H., Cao, J. X., and Shi, Q. X. (2008a) Integrated Quay crane and yard truck schedule for inbound containers, Proceedings of the IEEE International Conference on Industrial Engineering and Engineering Management (IEEM '08), 1219–1223, December 2008.

Lee, D.-H., Wang, H. Q., and Miao, L. (2008b) Quay crane scheduling with handling priority in port container terminals. Engineering Optimization. 40(2): 179–189.

Lee, D.-H., Wang, H. Q., and Miao, L. (2008c) Quay crane scheduling with non-interference constraints in port container terminals. Transportation Research Part E: Logistics and Transportation Review. 44(1): 124–135.

Legato, P., Canonaco, P., Mazza, R. M. (2009) Yard Crane Management by Simulation and Optimisation. Maritime Economics & Logistics. 11(1):36–57.

Legato, P., Truno, R., Meisel. F. (2012) Modeling and solving rich quay crane scheduling problems. Computers & Operations Research. 39(9): 2063–2078.

Levitin, G. and Abezgaouz, R. (2003) Optimal routing of multiple-load AGV subject to LIFO loading constraints. Computers & Operations Research. 30:397–410.

Li, W., Y. Wu, M. E. H. Petering, M. Goh, R. de Souza. (2009) Discrete time model and algorithms for container yard crane scheduling. European Journal of Operational Research. 198 (1):165–172.

Lim A (1998) The berth planning problem. Operations Research Letters. 22: 105–110.

Lim, A., Rodrigues, B., Xiao, F. and Zhu, Y. (2004) Crane scheduling with spatial constraints, Naval Research Logistics. 51: 386–406.

Lim, A., B. Rodrigues, X. Zhou. (2007) A m-parallel crane scheduling problem with a non-crossing constraint. Naval Research Logistics. 54(2): 115–127.

Linn, R., Liu, J.-y., Wan, Y.-w., Zhang, C., and Murty, K. G. (2003) Rubber tired gantry crane deployment for container yard operation. Computers & Industrial Engineering. 45(3): 429–442.

Linn, R. J., and Zhang, C.-Q. (2003) A heuristic for dynamic yard crane deployment in a container terminal. IIE Transactions. 35(2): 161–174.

Lu, Zhiqiang, Xiaole Han, Lifeng Xi, Alan L. Erera. (2012) A heuristic for the quay crane scheduling problem based on contiguous bay crane operations. Computers & Operations Research. 39(12): 2915–2928.

Meersmans, P. J. M. (2002) Optimization of Container Handling Systems, Tinbergen Institute Research Series, No. 271, ISBN 90–5170-591-3, http://repub.eur.nl/pub/1855/fewdis200203 13164327.pdf.

Meisel, F. (2011) The quay crane scheduling problem with time windows. Naval Research Logistics. 58(7): 619–636.
Meisel, F., C. Bierwirth. (2013) A framework for integrated berth allocation and crane operations planning in seaport container terminals. Transportation Science. 47(2): 131–147.
Mihalis M. Golias, M. Boile, Sotirios Theofanis (2009) Berth scheduling by customer service differentiation: A multi-objective approach. Transportation Research Part E. 45: 878–892.
Moccia, L., Cordeau, J. F., Gaudioso, M. and Laporte, G. (2006) A branch-and-cut algorithm for the quay crane scheduling problem in a container terminal. Naval Research Logistics. 53: 45–59.
Ng, W. C. (2005) Crane scheduling in container yards with inter-crane interference. European Journal of Operational Research. 164(1): 64–78.
Ng, W. C., and Mak, K. L. (2005a) An effective heuristic for scheduling a yard crane to handle jobs with different ready times. Engineering Optimization. 37(8): 867–877.
Ng, W. C., and Mak, K. L. (2005b) Yard crane scheduling in port container terminals. Applied Mathematical Modelling. 29(3): 263–276.
Ng, W. C., and Mak, K. L. (2006) Quay crane scheduling in container terminals. Engineering Optimization. 38(6): 723–737.
Nishimura E, Imai A, Papadimitriou S. (2001) Berth allocation planning in the public berth system by genetic algorithms. European Journal of Operational Research. 131: 282–292.
Nguyen, V. D., Kim, K. H. (2009) A dispatching method for automated lifting vehicles in automated port container terminals. Computers & Industrial Engineering. 56(3): 1002–1020.
Nishimura, E., Imai, A., and Papadimitriou, S. (2005) Yard trailer routing at a maritime container terminal. Transportation Research Part E. 41:53–76.
Park, T., R. Choe, S. M. Ok, K. R. Ryu. (2010) Real-time scheduling for twin RMGs in an automated container yard. OR Spectrum. 32: 593–615.
Park Y-M, Kim K. H. (2003) A scheduling method for berth and quay cranes. OR Spectrum. 25: 1–23.
Petering, M. E. H. (2010) Development and simulation analysis of real-time, dual-load yard truck control systems for seaport container transshipment terminals. OR Spectrum. 32: 633–661.
Petering, M. E. H., Wu, Y., Li, W., M. Goh, R. de Souza (2009) Development and simulation analysis of real-time yard crane control systems for seaport container transshipment terminals. OR Spectrum. 31(4): 801–836.
Peterkofsky R .I., Daganzo C. F. (1990) A branch and bound solution method for the crane scheduling problem. Transportation Research-B. 24B: 159–172.
PSA International (2006) "EMMA MÆRSK" - WORLD'S LARGEST CONTAINERSHIP CALLS AT PSA SINGAPORE TERMINALS, http://www.internationalpsa.com/psanews/pdf/nr061002.pdf. accessed August 12, 2014.
PSA International (2015) PSA CONTAINER THROUGHPUT PERFORMANCE FOR 2014, 19 January 2015, https://www.globalpsa.com/wp-content/uploads/nr150119.pdf. accessed March 26, 2015.
Salisbury, M. (2013) The JOC Top 50 World Container Ports, http://www.joc.com/port-news/joc-top-50-world-container-ports_20130815.html. accessed August 12, 2014.
Sammarra, M., Cordeau, J. F., Laporte, G. and Monaco, M. F. (2007) A tabu search heuristic for the quay crane scheduling problem. Journal of Scheduling. 10: 327–336.
Sharif, O., N. Huynh. (2012) Yard crane scheduling at container terminals: A comparative study of centralized and decentralized approaches. Maritime & Economics Logistics. 14(2): 139–161.
Sinriech, D. and Kotlarski, J. (2002) A dynamic scheduling algorithm for a multiple-load multiple-carrier system. International Journal of Production Research. 40:1065–1080.
Stahlbock R, Voß S. (2008) Operations research at container terminals: a literature update. OR Spectrum. 30(1):1–52.
Stahlbock, R., S. Voß. (2010) Efficiency considerations for sequencing and scheduling of double-rail-mounted gantry cranes at maritime container terminals. International Journal of Shipping and Transport Logistics. 2: 95–123.
Steenken, D., Voß, S., and Stahlbock, R. (2004) Container terminal operation and operations research — a classification and literature review. OR Spectrum. 26(1):3–49.

Stromberg, J. (2015). The MSC Oscar just became the world's biggest container ship. http://www.vox.com/2015/1/8/7513317/container-ship-msc-oscar. accessed March 26, 2015.

Tavakkoli-Moghaddam, R., Makui, A., Salahi, S., Bazzazi, M., and Taheri, F. (2009) An efficient algorithm for solving a new mathematical model for a quay crane scheduling problem in container ports. Computers & Industrial Engineering. 56(1), 241–248.

Tierney, K., S. Voß, and Stahlbock, R. (2014) A mathematical model of inter-terminal transportation, European Journal of Operational Research. 235(2): 448–460.

Vacca I, Bierlaire M, Salani M (2007) Optimization at container terminals: Status, trends and perspectives. In: Proceedings of the Swiss Transport Research Conference (STRC), Monte Verit´a/Ascona, pp 1–21, http://www.strc.ch/conferences/2007/2007_vacca.pdf. accessed August 25, 2014.

van der Meer, R. (2000) Operational Control of Internal Transport. TRAIL thesis series nr. t2000/5.

Vis, I. F. A., Carlo. H. J. (2010) Sequencing two cooperating automated stacking cranes in a container terminal. Transportation Science. 44(2): 169–182.

Vis, I. F. A. and de Koster, R. (2003) Transshipment of containers at a container terminal: An overview. European Journal of Operational Research. 147:1–16.

Vis, I. F. A. and Harika, I. (2004) Comparison of vehicle types at an automated container terminal. OR Spectrum. 26:117–143.

Xu, D., C. L. Li, J. Y. T. Leung. (2012) Berth allocation with time-dependent physical limitations on vessels. European Journal of Operational Research. 216(1) 47–56.

Yan, W., Huang, Y., Chang, D., and He, J. (2011) An investigation into knowledge-based yard crane scheduling for container terminals. Advanced Engineering Informatics. 25(3): 462–471.

Yang, C. H., Choi, Y. S., and Ha, T. Y. (2004) Simulation-based performance evaluation of transport vehicles at automated container terminals. OR Spectrum. 26:149–170.

Yuan, S., Skinner, B.T., Huang, S.D., Liu, D.K., Dissanayake, G., Lau, H. (2011) A job grouping approach for planning container transfers at automated seaport container terminals. Advanced Engineering Informatics. 25(3): 413–426.

Zhang, L., Khammuang, K., and Wirth, A. (2008) On-line scheduling with non-crossing constraints. Operations Research Letters. 36(5): 579–583.

Zhu, Y., and Lim, A. (2006). Crane scheduling with non-crossing constraint. The Journal of the Operational Research Society. 57(12): 1464–1471.

Chapter 3
Discrete Time Model and Algorithms for Container Yard Crane Scheduling

Container terminal (CT) operations are often bottlenecked by slow yard crane (YC) movements. Prime mover (PM) queues in front of the YCs are common. Hence efficient YC scheduling to reduce the PM waiting time is critical in increasing a CT's throughput. In this chapter, we develop an efficient model for YC scheduling by taking into account realistic operational constraints such as inter-crane interference, fixed YC separation distances, and simultaneous container storage/retrievals. Among them, only inter-crane interference has ever been considered in the literature. The model requires far fewer integer variables than the literature by using bi-index decision variables. We show how the model can be solved quickly using heuristics and rolling-horizon algorithm, yielding close to optimal solutions in seconds. The solution quality and solution time are both better than the literature even with additional constraints considered. The proposed formulations and algorithms can be extended to other problems with time windows and space constraints.

3.1 Introduction

Today, 90 % of the world's trade is transported via containers (Henwood 2006), mostly on containerships (Kim and Kim 1999; Steenken et al. 2004), leading to greater volume flows at the ports. Coupled with increased demand for speed to market, the CT operator faces constant pressure to reduce turnaround time and increase flow efficiency. Hence CT operators need better and efficient computational tools which can de-bottleneck a CT's container flow and increase container box throughput.

Vessels mooring at the berths wait for quay cranes (QCs) to upload/discharge their containers (see Fig. 3.1). The PMs shuttle between the QCs and YCs to move these boxes from the berth to the container yard (CY), and vice versa. Upon arrival at the blocks, the PMs queue in front of the YCs until they are served. As the typical YC to QC work rate is half (Ng and Mak 2005), YC operations is a potential

W. Li et al., *Planning and Scheduling for Maritime Container Yards*,
DOI 10.1007/978-3-319-17025-1_3

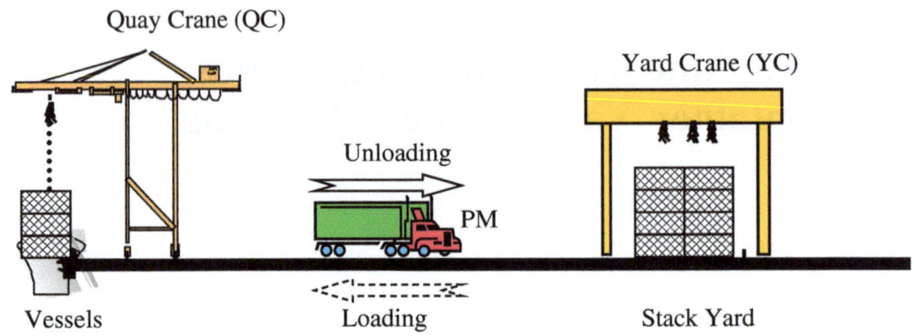

Fig. 3.1 Typical CT operations

bottleneck. The efficiency of yard operations depends heavily on the YCs' operations (Zhang et al. 2002).

The CY, a container storage area inside the CT of a port, for planning purposes, is typically partitioned as contiguous rectangular blocks (see Fig. 3.2). A grid is painted on the pavement indicating the *x–y* locations of the boxes. Bidirectional traffic lanes for the PMs occupy the space between the blocks. The blocks are divided along their length into 20-foot sections called slots. Each slot has several rows. Containers are stored alongside in each row and are stacked on top of each other. A typical block is six rows (6 × 8.5 ft) deep and forty slots (40 × 20 ft) long. The YCs are used to transfer containers in and out of the slots, straddling above the containers in each block and moving parallel to the length of the block. For a transshipment CT, with 40–60 slots per block, up to 60 moves are expected to be handled at each block (Ng 2005).

Generally, the terminal planners are informed of the vessel arrivals a few days prior. They then start to plan the storage/retrieval locations for the containers in the CY (Ng 2005). Hence the slot to be picked is already predetermined in most CTs.

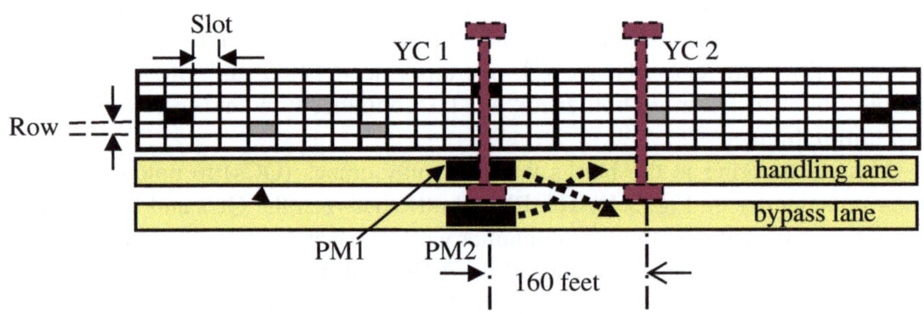

Fig. 3.2 160 ft separation distance between working YCs for a 28-slot block

However, due to the arrival uncertainty of the vessels, the container loading/unloading sequence in the QC work list is planned only several hours before the actual vessel arrival. This work list is then translated into an YC work list, using historical average container handling times for the QCs and YCs, and travel times for the PMs, respectively. The terminal schedulers then dispatch the YCs manually based on the actual container arrival pattern. Besides the import containers arriving from vessels, in a transshipment CT, there are export containers arriving from the landside of the terminal. Such export containers are stored at the stack yard temporarily and later moved to quayside by YCs and PMs (Froyland et al. 2008). This will also generate storage and retrieval moves for the YCs. Given this situation, it is optimal to generate an integrated YC work list which includes the work list translated from both the landside planning and quayside planning. However, in this chapter, for ease of exposition, only the YC work list translated from the QC work list is considered. Coordinating the movement of containers on the quayside is already a difficult problem in itself.

The gross crane rate (GCR) measures the average rate at which the QCs transfer containers between vessel and shore and is the most significant performance measure of a CT operation. A high GCR, while desirable, is however constrained by the ability of the PMs to reach the QCs, which in turn is constrained by the YCs' work rate. In short, even though a QC is technically capable of making 40 moves (load/discharge containers) per hour, the average QC rate at most container ports is currently far less (more than 30 %). An efficient YC schedule is therefore key.

Some unique operation constraints exist in YC scheduling. First, two YCs sharing the same bidirectional lane cannot cross gantry each other, i.e., an YC located at one position of a block cannot move to another position in the same block if another YC is in its path. Some studies (Ng 2005; Lim et al. 2007; Froyland et al. 2008) have considered this constraint. Second, the literature does not consider a safety distance between YCs. Currently, the safety distance between two YCs operating in the same lane must be at least 160 ft apart (Fig. 3.2). This greatly limits access to the containers in the intervening slots. For instance, if an YC is working at slot #8 in a block, then there can be no container handling in slots #1 to #15, effectively reducing the block workspace by 37.5 % for a 40-slot block. This safety distance is intended for the PMs to park and move between the YCs, and to move from the handling lane to a bypass lane and back. If YCs are closer than this, the PM (PM1 in Fig. 3.2) leaving the upstream YC (YC1 in Fig. 3.2) will not be able to pull out of the handling lane and onto the bypass lane through to the downstream YC (YC2 in Fig. 3.2). Third, in a transshipment CT, YCs often need to handle two kinds of moves, storage and retrieval, in a time window. Containers should either be stored or retrieved from CY by YCs from/to PMs (Fig. 3.1). Froyland et al. (2008) have considered storage and retrieval moves using heuristics to sequence container moves without setting a minimum YC separation distance. Other literature (Ng 2005; Ng and Mak 2005; Ng and Tsang 2005) only consider storage moves.

3.2 Literature Review

There are many papers on planning and scheduling problems in a CT. Vis et al. (2003), Steenken et al. (2004), and Stahlbock et al. (2008) provide excellent surveys of CT operations and research. For instance, Li et al. (2007) have developed models for allocating empty containers among ports to better manage and control their containers. Similarly, Kim et al. (2004) and Bish (2003) studied the scheduling for QCs. Many papers have considered the scheduling of a single YC where containers are grouped and the YC must retrieve containers from specified groups according to a fixed sequence, without due dates or release times, while minimizing travel distance. Since containers belonging to a specific group may be stored in multiple locations, both the YC route and the number of containers picked at each slot are decision variables (Kim and Kim 1999, 2003; Narasimhan and Palekar 2002).

Scheduling algorithms for multiple YCs are rarely addressed in the literature (Steenken et al. 2004; Stahlbock et al. 2008). Cheung et al. (2002) and Zhang et al. (2002) have developed methods for allocating YCs among the blocks in an entire CT and for scheduling cross-gantry moves. However, they do not generate detailed work schedules for the YCs and individual containers. Chen et al. (2007) have developed an integrated scheduling model for container handling where a Tabu search algorithm was used to solve the model. Again, the detailed work schedules for YCs were not generated. So far, most of the algorithms applied on the YC scheduling problem (Ng 2005; Ng and Mak 2005; Ng and Tsang 2005) only treat storage moves probably because it is easier to develop algorithms for storage moves. Recently, Ng (2005) has proposed a heuristic for constructing detailed work schedules for multiple YCs in a block with inter-crane interference, albeit without retrieval moves and minimum crane-distance separation.

3.2.1 Some Approaches to YC Work Schedules

Dispatching The YC scheduling problem is NP-hard (Ng 2005). Currently, dispatching is applied extensively in CTs using simple rules (e.g., SPT, EDD) for YC scheduling. However, such rules are myopic in nature (Hopp et al. 2000).

Simulation Another way to combat this NP-hard scheduling problem is through simulation, and generates schedules by applying different dispatching rules. However, finding an effective schedule is a time-consuming process and as dispatching rules are inherently myopic, simulation may not generate good schedules (Hopp et al. 2000).

Heuristics-based algorithms Some researchers (Ng 2005) apply heuristics-based algorithms for YC scheduling. These algorithms can generate a schedule quickly. However, lacking an efficient way to search for a globally good schedule, the final schedules generated by such algorithms may be far from optimal.

Metaheuristics-based approaches Another class of methods apply metaheuristics for YC scheduling. Recently, Ng and Mak (2005) proposed a branch-and-bound algorithm to solve the problem of storage moves, requiring about 444.7 s in average to solve YC scheduling problem with 25 jobs. Their algorithm cannot be applied on retrieval moves at the same time as the procedure fails to find the lower bound. Ng and Tsang (2005) later developed a genetic algorithm to solve a single YC scheduling problem. Again, only storage moves are considered.

MILP models for YC work schedules Ng (2005) proposed an MILP model as the benchmark for his heuristics-based algorithm considering inter-crane interference. Ng found that CPLEX took about 10 min to find the optimal schedule for a problem with 2 YCs and 10 jobs in a 40-slot block. Due to the exponential increase in solution time, solving MILP-based models directly can be intractable for real-sized problems. All research thus far use MILP-based models as benchmarks only.

This chapter develops an efficient model for YC work schedules by treating actual operation constraints such as inter-crane interference, fixed YC separation distances, and simultaneous storage/retrieval jobs handling. By detailed formulating and exploiting the structure of the problem, the model size is dramatically decreased. An effective heuristic is integrated into the model to further reduce the model size. We then apply a rolling-horizon algorithm to sequence jobs based on job target times. Jobs are scheduled at each iteration until all jobs are fixed. This allows unscheduled new jobs to be inserted among the existing set of jobs. The rolling-horizon algorithm can find near-optimal solutions. Thus far, the solutions of many scenarios tested are optimal. Since the number of jobs involved at each iteration is set to a small value (7 jobs/iteration is used in this chapter), the problem size of each iteration is small, reaching solution in less than one second usually. As the total problem size increases, the total solution time only increases linearly in N, i.e., $o\left(\frac{N}{\mathrm{JFPI}} * e^{\mathrm{JNPI}}\right)$, where N (>7) is the total number of jobs to be scheduled, JFPI is the number of jobs fixed at each iteration (2 is used in this chapter) and JNPI is the job number per iteration (7 is used here). The exponential term, e^{JNPI}, is much smaller compared to e^N.

3.3 The Model (DMIP1)

An MILP model is developed to handle the problem of scheduling single or multiple YCs in a yard block. Storage or retrieval moves arrive at the block with different target times. For retrieval jobs, the target time interval is the latest time interval during which a retrieval job can be handled and still meet the deadline set by the QCs. For storage jobs, the target time interval is the earliest time interval following the release of the job (i.e., the arrival of the corresponding PM) in the yard. In order to reduce the number of integer variables in our IP model, we discretize the time axis into 3.5-min intervals. Because an YC takes about 2–4 min to make a container transfer, we assume a 3-min handling time per container move.

Each container is scheduled to take place in exactly one interval and shall consist of 0.5 min of YC gantrying followed by 3 min of container handling. In the model, each YC handles at most one container per interval. In any given interval, all gantry moves start and end simultaneously and all container moves start and end simultaneously for all YCs.

3.3.1 Assumptions

The following assumptions are used in the model:

- Target times and locations of container moves are assumed known and fixed. The target times can be translated from the QC work list.
- The job handling time of an YC is usually 2–4 min (Ng and Mak 2005). We assume that the job handling time of all YCs is 3 min (i.e., 20 moves/h).
- 20–30 moves in a 2-h time window are used in the scenarios tested in this chapter. Each yard block contains 40–60 container slots.
- In a CT, YC–YC interference is usually more serious than PM–PM interference as YCs have more difficulty substituting for each other than PMs during actual operations (e.g., there are fewer YCs than PMs, YCs are much slower than PMs, and the minimum separation distance is larger for two YCs than for two PMs). Thus, more attention should be paid to scheduling YCs than to dispatching PMs. YCs also exhibit less variable travel and handling times than PMs, which means that they can follow a predetermined sequence of moves with a less variable result than PMs. Thus, generating YC schedules with the assumption that PMs are always available becomes very worthwhile.
- The minimum difference in slot numbers allowed for two YCs at the same time is assumed to be 8 slots (160 ft) in this chapter (SEP = 8).

3.3.2 Objective Function

The notations used for the indices, sets, parameters, and variables in the mathematical formulation are defined in Appendix 1. The objective is to minimize a linear combination of the retrieval earliness and storage and retrieval delays, namely,

$$\text{Minimize TC} = W_{\text{re}} \sum_{m \in R} \text{RE}_m + W_{\text{rd}} \sum_{m \in R} \text{RL}_m + W_{\text{sd}} \sum_{m \in S} \text{STL}_m \qquad \text{(obj)}$$

As a retrieval delay by the YC directly leads to the QC schedules and berth operations being delayed, and the storage delay only affects the yard operations, the weight for the total retrieval delay, W_{rd}, is set larger. Hence, we set $W_{\text{re}} = W_{\text{sd}} = 1$ and $W_{\text{rd}} = 2$.

3.3.3 Constraints

Each move takes place during exactly 1 time interval

$$\sum_{t \in T} X_{mt} = 1, \quad \forall m \in M \tag{3.1}$$

where T is the set of time intervals scheduled. It is determined by dividing the upper bound of the makespan of the optimal schedule by the interval length. The upper bound of the makespan (assumed to be a known value in Ng (2005)) is decided by trial-and-error. In this chapter, we set this upper bound to the maximum target time of all moves plus 50 min. That is, we reasonably assume that, in the optimal schedule, no move will be handled 50 min after the largest target time:

$$T = \{t: \ \text{Ord}_t * \text{IntLen} \ \leq \ \text{Max}H + 50\}$$

where $\text{Max}H = \text{Max}\{m \in M : \text{Tgt}_m\}$, Ord_t is the relative position of time interval t in set T and IntLen is the length of time interval.

Each move should be assigned to one crane

$$\sum_{c \in C} W_{mc} = 1, \quad \forall m \in M, \text{NC} > 1 \tag{3.2}$$

When the number of YCs is greater than one, constraint (3.2) ensures that each move is assigned to exactly one YC. If there is only one YC, the value of W_{mc} is fixed to 1,

$$W_{m,c1} = 1, \quad \forall m \in M, \ \text{NC} = 1$$

A storage move can take place no earlier than its target time interval

$$\sum_{t \in T} \text{IntLen} * (\text{Ord}_t - 1) * X_{mt} \geq \text{Tgt}_m, \quad \forall m \in S_m \tag{3.3}$$

At time *t*, at most NC YCs can work simultaneously

$$\sum_{m} X_{mt} \leq \text{NC}, \quad \forall t \in T \tag{3.4}$$

NC is the total number of YCs scheduled.

Computing the retrieval earliness, RE_m, of move *m*:

$$\text{RE}_m \geq \text{Tgt}_m - \sum_{t \in T} \text{IntLen} * (\text{Ord}_t - 1) * X_{mt}, \quad \forall m \in R_m \tag{3.5}$$

Computing the retrieval lateness, RL_m, of move *m*:

$$\mathrm{RL}_m \geq \sum_{t \in T} \mathrm{IntLen} * (\mathrm{Ord}_t - 1) * X_{mt} - \mathrm{Tgt}_m, \quad \forall m \in R_m \tag{3.6}$$

Computing the storage lateness, STL_m, of move *m*:

$$\mathrm{STL}_m \geq \sum_{t \in T} \mathrm{IntLen} * (\mathrm{Ord}_t - 1) * X_{mt} - \mathrm{Tgt}_m, \quad \forall m \in S_m \tag{3.7}$$

Identifying the status of YC *c* at time *t*

$$\mathrm{YY}_{mct} \geq X_{mt} + W_{mc} - 1, \quad \forall m \in M, \forall c \in C, \mathrm{NC} > 1, \forall t \in T \tag{3.8a}$$

$$\mathrm{YY}_{mct} \leq W_{mc}, \quad \forall m \in M, \ \forall c \in C, \mathrm{NC} > 1, \forall t \in T \tag{3.8b}$$

$$\mathrm{YY}_{mct} \leq X_{mt}, \quad \forall m \in M, \ \forall c \in C, \mathrm{NC} > 1, \forall t \in T \tag{3.8c}$$

Constraints (3.8a)–(3.8c) enforce that, if YC *c* is handling move *m* at time *t* (i.e., $X_{mt} = W_{mc} = 1$), YY_{mct} takes a value of 1. Note that constraints (3.8a)–(3.8c) ensure that the value of YY_{mct} is either 0 or 1. Thus, YY_{mct} can be defined as a continuous variable. Constraints (3.8a)–(3.8c) are only valid for multi-crane scenarios.

Moves handled by neighboring YCs

$$\sum_{n \in \mathrm{PLJ}_{mn}} \mathrm{YY}_{n,c+1,t} \leq 1 - \mathrm{YY}_{mct}, \quad \forall c \in C, \forall t \in T, \forall m \in M, \ c < \mathrm{NC}, \ \mathrm{NC} > 1 \tag{3.9}$$

Constraint (3.9) is only valid for multi-crane scenarios. It states that, if job *m* is being handled by YC *c* at time interval *t* (i.e., $\mathrm{YY}_{mct} = 1$), then YC $c + 1$ cannot handle moves which belong to PLJ_{mn}.

Without loss of generality, the slots of a block are ordered from left to right. We also assume that moves are ordered according to the slot number they are located in, i.e., if $m < n$, then $\mathrm{slot}_m \leq \mathrm{slot}_n$. With the above assumptions, we define set PLJ_{mn} as:

$$\mathrm{PLJ}_{mn} = \{m, n \in M : \ (\mathrm{slot}_n - \mathrm{slot}_m) < \mathrm{SEP} \wedge m \neq n\}$$

That is, those moves that are placed to the left of move m and those moves that are placed to the right of move *m* but their slot difference against move *m* is less than SEP belongs to PLJ_{mn} (Fig. 3.3). For example, in Fig. 3.3, YC 2 cannot handle moves that belong to PLJ_{mn} when YC 1 is handling move *m*. This is required by SEP apart of YCs and inter-crane interference.

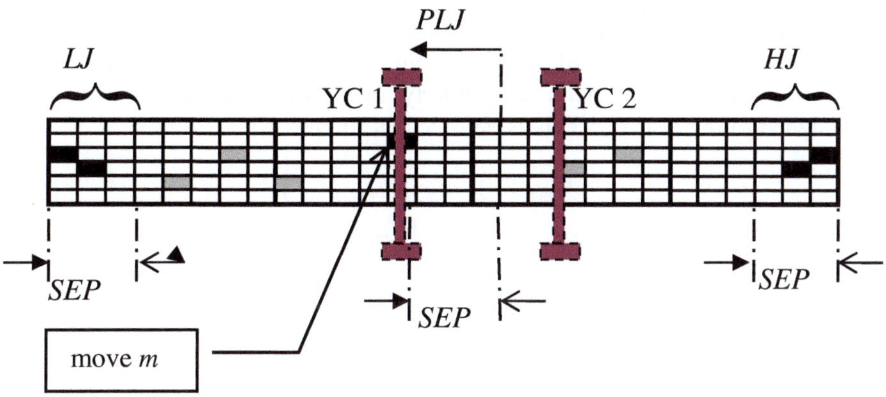

Fig. 3.3 Definitions of PLJ, HJ and LJ

At time *t*, an YC can handle at most one move

$$\sum_{m \in M} \mathrm{YY}_{mct} \leq 1, \quad \forall c \in C, \forall t \in T, \mathrm{NC} > 1 \tag{3.10}$$

Constraint (3.10) is only valid for multi-crane scenarios.

Consecutive moves handled by an YC

We do not allow any crane to gantry more than TVL (e.g. 8) slots between container handling operations. In other words, if YC c is handling move m at interval t (i.e., $\mathrm{YY}_{mct} = 1$) and the slot position of move n is too far (farther than TVL), then YC c cannot handle move n at interval $t + 1$ (i.e., $\mathrm{YY}_{nc,t+1}=0$). This is enforced by constraint (3.11) as follows:

$$\begin{aligned} &\mathrm{YY}_{n,c,t+1} \leq 1 - \mathrm{YY}_{mct}, \\ &\forall c \in C, \forall t \in T, t < \mathrm{NT}, \forall m, n \in \mathrm{FJ}_{mn}, \quad \mathrm{NC} > 1 \end{aligned} \tag{3.11}$$

NT is the total number of time intervals in set T. Constraint (3.11) is only valid for multi-crane scenarios. Set FJ_{mn} is defined as:

$$\mathrm{FJ}_{mn} = \{m, n \in M : \ |\mathrm{slot}_n - \mathrm{slot}_m| > \mathrm{TVL} \wedge m \neq n\}$$

That is, those moves that are TVL slots away from move m belong to FJ_{mn}. If there is only one YC, the following constraint is used to replace (3.11):

$$X_{n,t+1} \leq 1 - X_{mt}, \quad \forall t < \mathrm{NT}, \forall m, n \in \mathrm{FJ}_{mn}, \ \mathrm{NC} = 1 \tag{3.12}$$

Moves located at the first and last SEP slots

Moves located at the first SEP slots (see Fig. 3.3) can only be assigned to the first YC due to inter-crane interference. Similarly, moves at the last SEP slots can only be assigned to the last YC. This is enforced as follows:

$$W_{mc} = 1, \forall m \in \mathrm{LJ}_m, c = 1$$
$$W_{mc} = 1, \forall m \in \mathrm{HJ}_m, c = \mathrm{NC}$$

Set LJ_m is defined as:

$$\mathrm{LJ}_m = \{m \in M : \ \mathrm{slot}_m \leq \mathrm{SEP}\}$$

LJ_m includes moves located at the first SEP slots (see Fig. 3.3). Similarly, HJ_m is defined as:

$$\mathrm{HJ}_m = \{m \in M : \ \mathrm{slot}_m > (\mathrm{NSL} - \mathrm{SEP})\}$$

where NSL is the total number of slots. HJ_m includes moves located at the last SEP slots (see Fig. 3.3).

As the storage moves can only be handled after their target times, we can fix some X_{mt} to 0:

$$X_{mt} = 0, \forall m, t \in \mathrm{XS}_{mt}$$

where set XS_{mt} denotes time intervals t which are earlier than the target time of storage move m. It is defined as:

$$\mathrm{XS}_{mt} = \{m \in S_m, t \in T : \ \mathrm{IntLen} * (\mathrm{Ord}_t - 1) \ - \ \mathrm{Tgt}_m < 0\}$$

3.3.4 Results of DMIP1

The number of YCs in each block varies from 1 to 4. Thirty test scenarios with 20–32 moves in 40-slot blocks and thirty test scenarios with 20–32 moves in 60-slot blocks are randomly generated. ILOG CPLEX 9.0 is used to solve all the models developed in this chapter on a Pentium 1.6 GHz computer. The MILP relative optimality gap of CPLEX is set to 0 (optcr = 0.0). No priority branching for the binary variables is used (PriorOpt = 0). All other settings follow directly from system default.

An illustrative example (EX1), which contains 2 YCs, 60 slots and 32 jobs, is used to generate the results for DMIP1. Table 3.1 contains the dataset for EX1. The start time of the scheduling horizon is 168 min. Using the formulations described in Sects. 3.3.2 and 3.3.3, DMIP1 involves 1291 binary variables (49 time intervals),

Table 3.1 Data for the illustrative example (EX1)

Moves	Target time (min)	$Slot_m$	Type[a]
m1	270.56	2	S
m2	261.317	3	S
m3	226.336	6	S
m4	198.957	8	S
m5	168.451	9	S
m6	214.004	11	S
m7	272.078	13	S
m8	217.108	15	S
m9	200.291	16	S
m10	250.959	20	S
m11	171.535	22	S
m12	204.811	23	S
m13	172.899	24	S
m14	260.132	26	S
m15	204.783	27	S
m16	262.883	31	S
m17	192.807	32	S
m18	261.393	33	S
m19	212.644	36	S
m20	261.324	37	S
m21	290.849	37	R
m22	220.24	42	R
m23	246.347	43	R
m24	174.536	44	R
m25	230.677	47	R
m26	207.904	47	R
m27	217.201	51	R
m28	290.617	52	R
m29	221.771	54	R
m30	180.469	58	R
m31	204.882	59	R
m32	224.63	60	R

[a]*R* Retrieval move; *S* Storage move

4813 single variables, and 84,212 single equations. The solution time is 8604.7 s. The total PM waiting time is 106.647 min.

We note that in the MILP model developed in Ng (2005), three tri-index binary variables, X_{mnc}, Y_{cSLt}, W_{mct} are defined which involves 32 * 32 * 2 + 2 * 60 * 49 + 32 * 2 * 49 = 11,064 binary variables in a 49-time interval horizon. Furthermore, to increase the accuracy, each interval is defined to be the time YCs required to travel a slot in Ng's model. On average, an YC takes 4 s to gantry one slot. Thus, in a 2-h time window, there are 1800 time intervals. The number of

Table 3.2 Results of DMIP1

Scenario	Solution time (s)		
	Mean	Maximum	Minimum
40 slots	1917.5	50,208.2	0.6
60 slots	7007.3	146,374.7	2.5
Pure storage moves	450.5	4815.7	0.6
Storage + Retrieval	7137.1	146,374.7	0.7

binary variables involved in Ng's model is now 32 * 32 * 2 + 2 * 60 * 1800 + 32 * 2 * 1800 = 333,248, which is considerably larger than DMIP1.

The results of DMIP1 are shown in Table 3.2. Due to the isolated cases of extremely large maximum solution times, the mean solution time to solve 40-slot blocks with 2 YCs is 1917.5 s. For 60-slot blocks, the mean solution time is 7007.3 s. DMIP1 can yield a good solution in 450.5 s on average for scenarios with pure storage moves. This is better than the average of 10 min to solve smaller problems with only 10 pure storage jobs in Ng (2005). Note that the solution time may increase exponentially when the number of jobs increases from 10 to 20–32. Generally, DMIP1 takes much longer time (about 7137.1 s) to obtain a good solution when retrieval moves are involved in the scenarios.

3.4 Model Integrated with Heuristics (DMIP2)

Though the number of binary variables has been reduced significantly by defining only two bi-index binary variables, W_{mc} and X_{mt}, in DMIP1 compared to defining tri-index binary variables (see Ng 2005), the size of the model still increases significantly as the number of moves increases. We propose a heuristic to narrow the search space in DMIP2. We observe that, for both storage and retrieval moves, the larger the difference between the job finish time of a move and its target time, the larger the total PM waiting time. Since the total PM waiting time is to be minimized, the job finish time is arranged around its corresponding target time as near as possible in an optimal solution. We can reasonably assume that the job finish time of each move is placed inside a certain range (job handling range) around its target time. To apply the above heuristic into the model, we first define set TS_{mt} for storage moves as follows:

$$\text{TS}_{mt} = \{m \in S_m, t \in T : \text{IntLen} * (\text{Ord}_t - 1) - \text{Tgt}_m \geq 0 \cap \text{IntLen} * (\text{Ord}_t - 1) - \text{Tgt}_m < \text{SU} * \text{IntLen}\}$$

That is, the time intervals between the target time and SU * IntLen above the target time of move m belongs to set TS_{mt}. Similarly, we define set TR_{mt} for retrieval moves and TSR_{mt} for both of them.

$$\mathrm{TR}_{mt} = \{m \in R_m, t \in T : \mathrm{Tgt}_m - \mathrm{IntLen} * (\mathrm{Ord}_t - 1) \geq -\mathrm{RL} * \mathrm{IntLen} \cap \mathrm{Tgt}_m - \mathrm{IntLen} * (\mathrm{Ord}_t - 1) < \mathrm{RU} * \mathrm{IntLen}\}$$

$$\mathrm{TSR}_{mt} = \mathrm{TS}_{mt} \cup \mathrm{TR}_{mt}$$

We force move m to be handled within TSR_{mt} by:

$$X_{mt} = 0, \forall m, t \notin \mathrm{TSR}_{mt}; \; \mathrm{YY}_{mct} = 0, \forall m, t \notin \mathrm{TSR}_{mt}$$

SU, RL, and RU are parameters used to define the job handling range. Their values can be adjusted according to the density of the planned job arrivals. If a lot of moves arrive within a short period of time, the YCs will be quite busy. Hence it is possible that some moves may be scheduled far from their target times. We set the values of SU, RL, and RU higher in this case. SU, RL, and RU are all set to value of 8.0, which is conservative and satisfies all the scenarios tested in this chapter. Some constraints should be modified when applied to set TSR_{mt}. For example, constraint (3.9) is defined within TSR_{mt}:

$$\sum_{n \in \mathrm{PLJ}_{mn}} \mathrm{YY}_{n,c+1,t} \leq 1 - \mathrm{YY}_{mct}, \quad \forall c \in C, \forall t \in T, \forall m \in M, \; c < \mathrm{NC}, \; \mathrm{NC} > 1, \; \forall m, t \in \mathrm{TSR}_{mt}. \tag{3.9'}$$

Similarly, constraints (3.8a)–(3.8c), (3.10), (3.11), (3.12) should also be defined within TSR_{mt}.

3.4.1 Results of DMIP2

For EX1 described in Sect. 3.3.4, DMIP2 involves 398 binary variables, 4548 single variables, and 30,572 single equations. The model size is significantly reduced compared to DMIP1. EX1 was solved in 676.5 s with a total PM waiting time of 106.647 min.

In Table 3.3, all scenarios are solved to optimality by setting the MIP relative optimality gap to zero. The mean solution time needed to solve 40-slot blocks is

Table 3.3 Results of DMIP2

Scenario	Solution time (s)			Gap (%)
	Mean	Maximum	Minimum	$(\mathrm{TCT} - \mathrm{TCT}_{\mathrm{LB}})/\mathrm{TCT}_{\mathrm{LB}}$
40 slots	208.3	5390.6	0.2	NA
60 slots	1703.5	36,842.6	0.4	NA
Pure storage moves	51.9	924.0	0.2	5.1
Storage + Retrieval	1558.6	36,842.6	0.3	NA

208.3 s. For 60-slot blocks, the mean solution time is 1703.5 s. Generally, scenarios can be solved in several seconds. The average solution time is lengthened by 2–3 scenarios with very long solution times. To compare the results with the literature, we calculate the lower bounds (TCT_{LB} in Table 3.3) for the scenarios with pure storage moves by applying the evaluation procedure proposed by Ng (2005). The average/maximum/minimum TCT gaps for pure storage scenarios from our model are 5.1 %/14.4 %/2.4 %. That is, even with additional constraints (YC safety distance, simultaneous storage/retrieval moves) considered in our chapter, the solution quality is still better than the current literature (7.3 % gap achieved in Ng 2005).

We note that DMIP2 generally takes a much longer time (average of 1558.6 s) to obtain a good solution when retrieval moves are involved while DMIP2 can yield a good solution in about 51.9 s for scenarios with pure storage moves, making it much more efficient than the model developed by Ng (2005). This may be because retrieval moves can be either handled before and after their target times, while storage moves can only be handled after.

3.5 Rolling-Horizon Algorithm for DMIP2 (DMIP3)

Though DMIP2 has reduced the model size and the solution time significantly, it is unstable and sometimes still requires very long solution time for some scenarios. As observed in Sect. 3.4, the job finish time of a move should be scheduled as near as possible around its target time. In other words, a move *n* which arrives much later than move *m* is very likely to finish later than move *m*, otherwise a large gap between the job finish time and target time will be incurred. This gives us the opportunity to divide the jobs into many groups. Moves with near job target times are grouped together and their handling sequence can be interchanged. All moves in a higher group (moves with bigger job target times) will be handled later than a lower group (moves with smaller job target times). For example, in Fig. 3.4, all scheduled job handling times of moves in group H are bigger than moves in group M and L. Thus, we can divide the problem into several subproblems; each subproblem involves moves of only one group. Because each subproblem involving fewer moves can be solved quickly, the entire YC scheduling problem can be solved quickly.

In Fig. 3.4, we observe that, the moves in group M which have big target times and moves in group H which have small target times are actually near jobs. Their handling sequence can possibly also be interchanged. Thus, when implementing the algorithm, all subproblems are generated in a rolling-horizon fashion. In Fig. 3.5, the first subproblem includes moves *m*1 to *m*5. After solving this small subproblem, two moves (*m*1 and *m*2) with the smallest handling times are fixed. In the second subproblem, moves *m*3 to *m*5 are also included because their handling sequence may possibly be interchanged with the two newly added moves (*m*6 and *m*7). In subproblem 2, if several retrieval moves have very near target times, it is possible to insert one or several of these moves into the empty space among the jobs fixed.

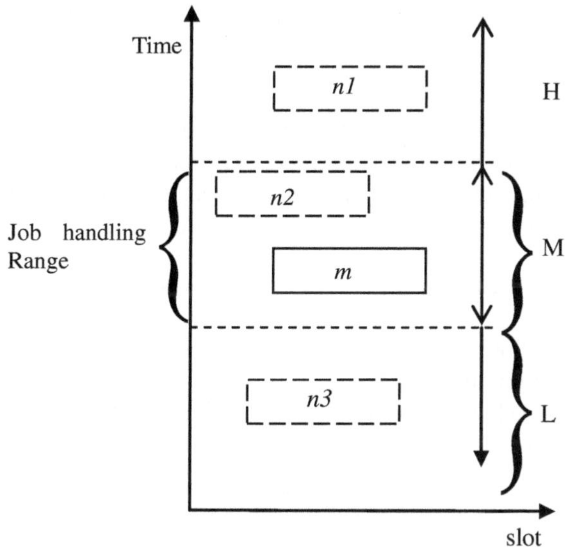

Fig. 3.4 Definition of neighboring moves

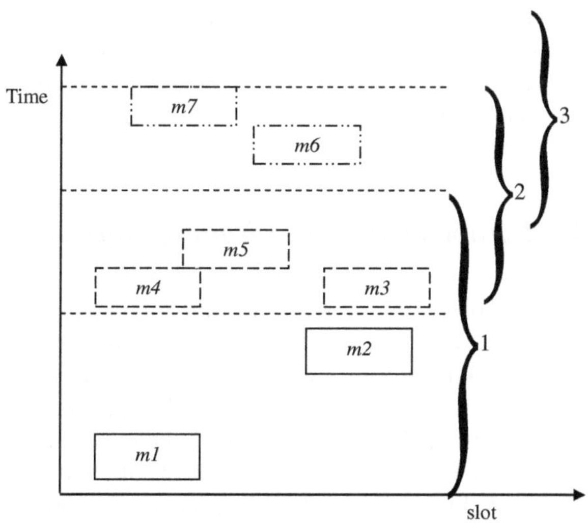

Fig. 3.5 Definition of rolling subproblems

For example, if moves *m*3 to *m*5 are retrieval jobs and their target times are almost the same, we may need to insert one of these moves into the space between moves *m*1 and *m*2 which have been fixed in subproblem 1. After solving subproblem 2, we can fix two more moves and add two new moves into the subproblem 3. The algorithm stops when all moves have been fixed.

3.5.1 Constraints and Sets Used in the Rolling-Horizon Algorithm

To implement the rolling-horizon algorithm, some sets and constraints in Sects. 3.3 and 3.4 need to be modified. We first define set PM_m to include moves involved in a subproblem. Then we redefine sets FJ_{mn}, PLJ_{mn}, HJ_{mn}, LJ_{mn} and TSR_{mt} under set PM_m. For example, set FJ_{mn} is redefined as:

$$\mathrm{FJ}_{mn} = \{m, n \in M : |\mathrm{slot}_n - \mathrm{slot}_m| > \mathrm{TVL} \wedge m \neq n \wedge \mathrm{PM}_m \wedge \mathrm{PM}_n\}$$

Note that in the above definition, only "^PM_m^PM_n" is added from the previous definition.

Related constraints should be modified under set PM_m in DMIP3. For example, constraint (3.2) is replaced by (3.2′):

$$\sum_{c \in C} W_{mc} = 1, \quad \forall m \in \mathrm{PM}_m, NC > 1 \tag{3.2′}$$

We allow unscheduled free jobs (i.e., PM_m) to be inserted among the existing set of fixed jobs (i.e., FxJ_m) in the rolling-horizon algorithm. Several constraints should also be added to define the relationship between free moves and fixed moves. These constraints are similar to (3.4), (3.9)–(3.12) except that they are defined under sets PM_m and FxJ_m.

3.5.2 Flow Diagram of Rolling-Horizon Algorithm

Figure 3.6 shows the flow diagram of the rolling-horizon algorithm. First, the *include* files are read to define the whole problem (number of moves, target time, job type, slot position, YC number, etc.). Then we set the value of parameters JNPI (=7) and SU, RL, RU (=8). JNPI can be adjusted according to the tradeoff between the solution quality and time. The solution quality increases as JNPI, the solution time of each iteration and hence the total solution time may increase. *Sets* FxJ_m, TsFin_m, WFin_{mc} XFin_{mt}, YYFin_{mct} are set to empty at the beginning. All moves are sorted according to their target times and saved in set Order_m. Set PM_m includes moves involved in an iteration. At the beginning, PM_m includes the first JNPI moves with the smallest target times. With PM_m defined, sets FJ_{mn}, PLJ_{mn}, HJ_{mn}, LJ_{mn} and TSR_{mt} are updated. DMIP3 is then solved to obtain the schedule for the first JNPI moves. This usually takes less than 0.1 s to obtain an optimal solution because the size of DMIP3 is very small at each iteration. Two moves with the smallest job handling times are fixed and added into the set FxJ_m from the result of DMIP3. Their job finishing times and handling YCs are recorded into sets TsFin_m and WFin_{mc}, respectively. The RE (retrieval earliness), RL (retrieval lateness), STL

Fig. 3.6 Rolling-horizon algorithm

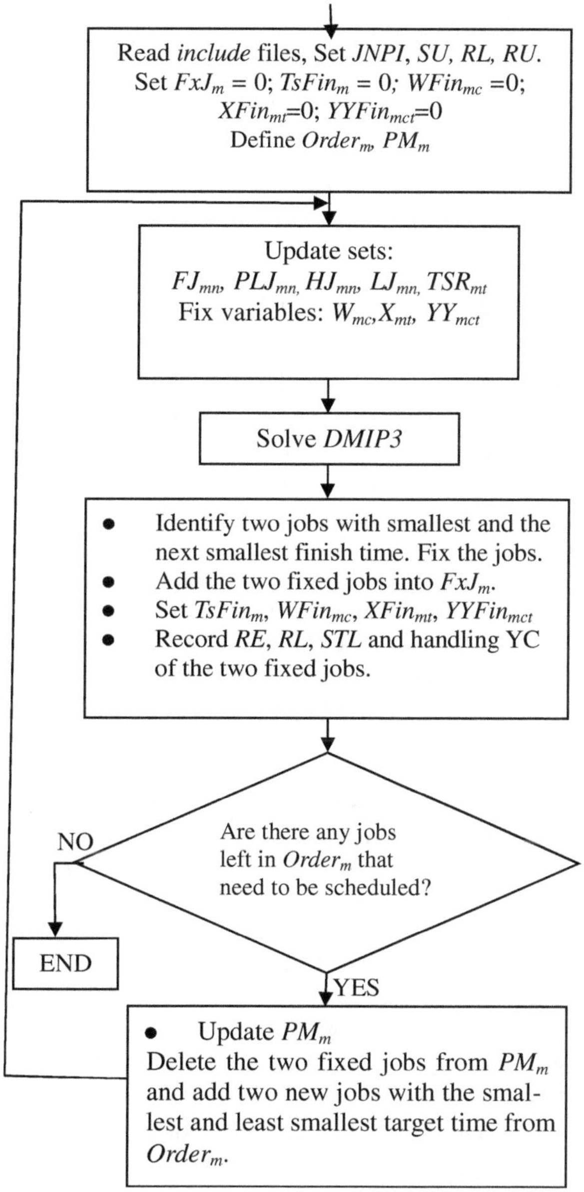

(storage lateness) are also recorded. Next, we check whether there are moves left in the set Order_m. If NO, the algorithm terminates. Otherwise, the set PM_m is updated by deleting moves fixed and adding new moves from Order_m. With updated PM_m, sets NJ_{mn}, PLJ_{mn}, etc., are updated accordingly and the algorithm continues until all moves in Order_m are scheduled.

Table 3.4 Results of DMIP3

Scenario	Solution time (s)			Gap (%)
	Mean	Maximum	Minimum	$(TCT - TCT_{LB})/TCT_{LB}$
40 slots	3.6	9.5	0.7	NA
60 slots	4.3	7.5	1.6	NA
Pure storage moves	3.0	6.4	0.7	5.2
Storage + Retrieval	4.6	9.5	1.7	NA

3.5.3 Results of DMIP3

The results of the model with rolling-horizon algorithm are shown in Table 3.4. For EX1 described in Sect. 3.3.4, DMIP3 involves 74 binary variables, 617 single variables, and 1606 single equations at the first iteration. DMIP3 obtains the solution for EX1 in 3.1 s with a total PM waiting time of 106.647 min which is also optimal. DMIP3 obtains a near-optimal solution in 3.9 s on average for all scenarios tested. This is much faster than the solution time (444.72 s on average) required in the paper of Ng and Mak (2005) solving for 25 moves in a block. DMIP3 is very robust because the maximum solution time is only 9.5 s. By applying the lower bound evaluation procedure proposed by Ng (2005), the average/maximum/minimum TCT gaps for pure storage scenarios are 5.2 %/16.6 %/3.0 %. Again, this is smaller than the current literature. Because TCT_{LB} is usually 1–3 % lower than the optimal TCT, the actual average TCT gap for pure storage scenarios is less than 5.2 %. A noticeable feature of DMIP3 is that it finds optimal solutions for all scenarios with pure storage moves in only 3.0 s on average. From the 60 scenarios tested, the average/maximum/minimum ratio of the optimal DMIP3 objective to the optimal DMIP1 objective is 2.6 %/26.2 %/0.0 %. We note that DMIP3 obtained optimal solutions for 49 out of the 60 tested scenarios.

Although the decrease in size is accompanied by an increase in the number of subproblems to solve, the total solution time still reduces dramatically. The results suggest that it is much better to solve a larger number of subproblems than to solve the MILP model as a whole. As the number of jobs involved at each iteration is set to a small value, the model size of each iteration is very small. As the total problem size increases, the total solution time only increases linearly in N, i.e., $o\left(\frac{N}{\mathrm{JFPI}} * e^{\mathrm{JNPI}}\right)$. With JNPI much smaller than N, the exponential term, e^{JNPI}, is much smaller compared to e^N.

3.6 Influence of Weights in Objective Function

The influence of the weight used for the total retrieval delay, W_{rd}, on PMWT, RE, RL, and STL is analyzed. Figures 3.7 and 3.8 show the results of PMWT, RE, RL, and STL at different W_{rd} for two scenarios (scenarios WI and WII). In general, RL

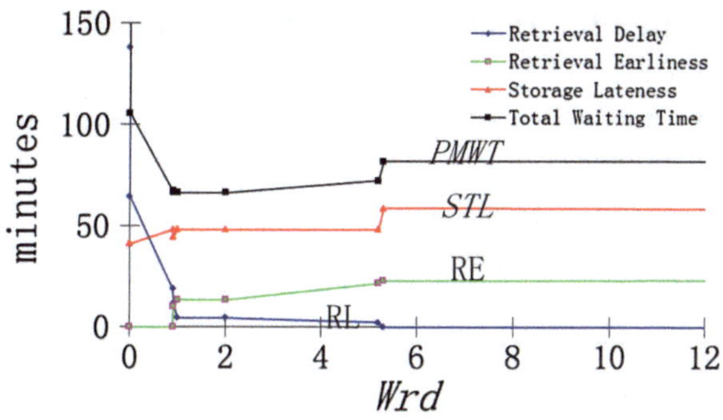

Fig. 3.7 Influences of W_{rd} for scenario WI

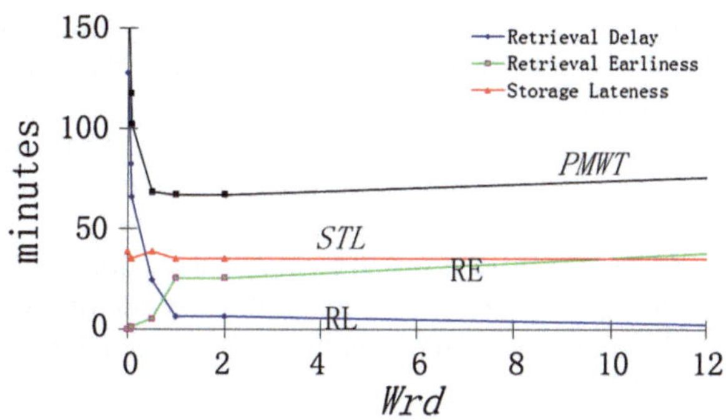

Fig. 3.8 Influences of W_{rd} for scenario WII

decreases to zero as W_{rd} increases. This is reasonable because a large RL results in a large objective value. The PMWT decreases as W_{rd} increases initially and then increases again. This is because, when W_{rd} is small, large RL occurs and PMWT is big. As W_{rd} increases, RL decreases quickly and hence PMWT decreases. When W_{rd} is very large, RL tends to zero. The schedule has to introduce large STL and *RE* to compensate. This increases the PMWT again.

3.7 Conclusion

This chapter develops efficient discrete time models for container YC work schedules. By applying heuristics and a rolling-horizon algorithm, we show that the model size is greatly reduced systematically and the solution time is shortened from days to seconds. The algorithm yields higher solution quality in a very short time compared to other heuristics used in the literature. The proposed formulations and algorithms can be extended to problems with time windows and space constraints such as quay crane scheduling or berth allocation (Wang et al. 2007). For future research, we suggest developing robust YC scheduling models by taking into account uncertain PM arrival times. Another future research development can consider the effect of landside conditions.

Appendix 1: Definition of Sets and Parameters

The notations used in the mathematical formulation are as follows.

(a) Indices

c, c' = yard cranes, c, c' = 1, 2,…,C
m, n = container moves, m, n = 1, 2,…,M
SL = slot number, SL = 1, 2,…,TSL
t = time interval

(b) Sets

C = number of YCs working in the block
FJ_{mn} = moves n that are TVL slots away from move m
FxJ_{m} = jobs that have been fixed after the current iteration and all previous iterations
HJ_{m} = includes moves located at the last SEP slots in the block
LJ_{m} = includes moves located at the first SEP slots in the block
M = set of container moves to be scheduled
Order_{m} = all jobs sorted by their target times
PLJ_{mn} = moves that locate on the left side of move m and moves that locate at the right side of move m but their slot difference against move m are less than SEP
PM_{m} = moves involved in a subproblem of rolling-horizon algorithm
R_m = set of yard retrieval moves
S_m = set of yard storage moves (the total number of items in R_m and S_m is M)
T = set of time intervals scheduled
TSL = number of slots in the block
XS_{mt} = time intervals t which are earlier than the target time of storage move m
TSR_{mt} = time intervals t which belong to job handling range of storage or retrieval move

(c) Parameters

IntLen = length of time interval
SU, RU, RL = parameter used to defined the job handling range
JNPI = job number per iteration
Max$H$ = the maximum target time of moves
NC = total number of YCs scheduled
NSL = total number of slots in the block
NT = total number of time intervals in set T
SEP = minimum difference in slot numbers allowed for two YCs at the same time. SEP is assumed to be 8 slots in this chapter
Slot_{m} = Slot number where move m takes place. Without loss of generality, we assume that if $m < n$, then $\mathrm{slot}_m \le \mathrm{slot}_n$
TVL = Maximum number of slots an YC can gantry in half minute (8 is used here)

Tgt_m = target time for move m in the block. For retrieval moves, this is the latest job start time that meets the deadline set by the QCs. For storage moves, this is the earliest job start time following the release of the job in the block
$TsFin_m$ = Job start time (Ts_m) of moves that have been fixed
W_{re} = weight assigned to total retrieval earliness in the objective function
W_{sd} = weight assigned to total storage delay in the objective function
W_{rd} = weight assigned to total retrieval delay in the objective function
$WFin_{mc}$ = W_{mc} of moves that have been fixed
$XFin_{mt}$ = X_{mt} of moves that have been fixed
$YYFin_{mct}$ = YY_{mct} of moves that have been fixed

(d) Variables

RE_m = amount of retrieval earliness for move m
RL_m = amount of retrieval lateness for move m
STL_m = amount of storage lateness for move m
TC = the linear combination of retrieval earliness and storage and retrieval delays. It is the objective value to be minimized.
W_{mc} = 0–1, binary variable to denote if container move m is assigned to YC c
X_{mt} = 0–1 variable to denote if move m is scheduled to take place during time interval t
YY_{mct} = 0–1, continuous variable to denote if YC c handles moves m at time interval t

References

Bish E. K. A (2003) multiple-crane-constrained scheduling problem in a container terminal. *European Journal of Operational Research.* 144; 83–107.
Chen L., Bostel N., Dejax P., Cai J., Xi L. (2007) A tabu search algorithm for the integrated scheduling problem of container handling systems in a maritime terminal. European Journal of Operational Research. 181: 40–58.
Cheung R. K., Li C., and Lin W. (2002) Interblock crane deployment in container terminals. Transportation Science. 36:79–93.
Froyland G.., Koch T., Megow N., Duane E., Wren H. (2008) Optimizing the landside operation of a container terminal. OR Spectrum. 30:53–75.
Henwood Rachel (2006) The practitioner's definitive guide: seafreight forwarding, SNP Reference.
Hopp W. J., Spearman M. L. (2000) Factory Physics: foundations of manufacturing management, Irwin/McGraw-Hill.
Kim K. H. and Kim K. Y. (1999) An optimal routing algorithm for a transfer crane in port container terminals. Transportation Science. 33:17–33.
Kim K. Y. and Kim K. H. (2003) Heuristic algorithms for routing yard-side equipment for minimizing loading times in container terminals. Naval Research Logistics. 50:498–514.
Kim K. H., Park Y. (2004) A crane scheduling method for port container terminals. European Journal of Operational Research. 156:752–768.
Lim A., Rodrigues B., Xu Z. (2007) A m-Parallel Crane Scheduling Problem with a Non-crossing Constraint. Naval Research Logistics. 54:115–127.

Li J., Leung S., Wu Y., Liu K. (2007) Allocation of empty containers between multi-ports. European Journal of Operational Research. 182:400–412.

Narasimhan A. and Palekar U. S. (2002) Analysis and algorithms for the transtainer routing problem in container port operations. Transportation Science. 36:63–78.

Ng, W. C.(2005) Crane scheduling in container yards with inter-crane interference. European Journal of Operational Research. 164:64–78.

Ng W. C. and Mak, K. L. (2005) Yard crane scheduling in port container terminals. Applied Mathematical Modelling. 29:263–275.

Ng W. C. and Tsang, W. S. (2005) Scheduling yard crane in a port container terminal using genetic algorithm. The First International Conference on Transportation Logistics (T-Log 2005), 27–29 July 2005; Singapore.

Stahlbock R., Voß S. (2008) Operations research at container terminals: a literature update. OR Spectrum. 30:1–52.

Steenken D., Voß S., Stahlbock R. (2004) Container terminal operation and operations research—a classification and literature review. OR Spectrum. 26:3–49.

Vis I. F. A., Koster R.D. (2003) Transshipment of containers at a container terminal: an overview. European Journal of Operational Research.147:1–16.

Wang F., Lim A. (2007) A stochastic beam search for the berth allocation problem. Decision Support Systems. 42:2186–2196.

Zhang C., Wan Y., Liu J., and Linn R. J. (2002) Dynamic crane deployment in container storage yards. Transportation Research B. 36:537–555.

Chapter 4
A Continuous-Time Model for Multiple Yard Crane Scheduling with Last-Minute Job Arrivals

Container terminal (CT) operations are often bottlenecked by slow yard crane (YC) movements. Efficient YC scheduling to reduce the PM waiting time is therefore critical in increasing a CT's throughput. This chapter develops an efficient continuous-time MILP model for YC scheduling. The model treats realistic operational constraints such as multiple intercrane interference, fixed YC separation distances, simultaneous container storage/retrievals, realistic YC acceleration/deceleration stages and gantry time, and require far fewer integer variables than previous work. The model significantly improves the solution quality compared to the existing discrete-time models and other heuristics found in the literature. Using heuristics and a rolling-horizon algorithm (RHA), our model can solve actual container yard (CY) problems quickly and robustly in polynomial time. Also, to cope with the last-minute container arrivals which can disrupt routine CT operations, two methods for handling these last-minute job insertions are discussed and compared.

4.1 Introduction

4.1.1 Background

Increased reliance on maritime trade within Asia necessitates growth in containers, carriers, CT, and other container port-related equipment such as the yard and quay cranes and prime movers. A recent industry report mentioned that intra-Asian trade (excluding the Middle East, Indian subcontinent, and Australasia) was 28.6 million TEU in 2007, or equivalent to 20 % of the total global trade (Eyefortransport 2008). This excludes any business moving within its confines on a feeder basis which is bound to markets in Europe, the US, and South America. This volume is expected to reach 50.7 million TEU by 2013. Any operational efficiency is thus constrained by the bottlenecks residing at various points of the maritime supply chain,

W. Li et al., *Planning and Scheduling for Maritime Container Yards*,
DOI 10.1007/978-3-319-17025-1_4

particularly at the landside operations. With greater expected container volumes and port calls, some of the major ports in Asia are already operating near or exceeding their capacity such as India (Froyland et al. 2008). As nodal players in the supply chain, CT operators face an imperative to increase container flow efficiency through the CY. Sound port planning and smooth execution of the equipment flow within the CY is thus a competitive advantage for success. However, practical constraints limit the CY from attaining its desired efficiency. Slower than desired YC movements and last-minute container arrivals into a CY often impede the performance of CT operations.

We now examine some of these operating constraints and highlight how the literature has attempted to reconcile these constraints in their models. First, in a work safe environment, regulatory compliance dictate that no two YCs traversing on an identical bidirectional lane can cross gantry each other. This is to reduce the possibility of YC–YC collisions, which can further hamper the throughput rate. This constraint was first studied by Lim et al. (2004) and has been later well examined and modeled in Chap. 3 and the literature (e.g., Ng 2005; Zhu and Lim 2006; Lim et al. 2007; Cao et al. 2008; Froyland et al. 2008; Wen et al. 2010). Second, to prevent tailgating, some ports impose a minimum safety distance of 160 ft between any two YCs. This constraint affects container pick efficiency as it delays and sometimes limits access to the containers residing in the intervening slots. Chapter 3 has discussed this issue. Third, in Asian CTs such as Singapore and Hong Kong, due to the constraint of space and for better land use, the containers for storage and retrieval are normally kept together. This naturally lengthens the retrieval time (including search time) and thus requires more planning operationally. A reasonable approach to improving the crane utilization is to minimize the deadhead time in these CTs. Put simply, it is better for YC to be able to put a container in its rightful slot position in the yard and then pick up another container almost immediately, to minimize unnecessary travel time within the yard. However, to handle the "simultaneous" drop and pick of containers requires much advanced planning. For expediency, yard operators tend to limit the number of simultaneous pick and drop moves in favor of the trade-off for idle YCs. Most of the literature (e.g., Ng 2005; Ng and Mak 2005) consider only storage moves. This chapter develops an approach for generating realistic YC work schedules, taking into account the above operating constraints, i.e., intercrane interferences, fixed YC separation distances, and simultaneous storage/retrieval jobs handling.

4.1.2 Literature Review

Simulation models for CT operations in the literature address various decision problems (Vis 2006; Petering et al. 2009). For instance, Vis (2006) compared the performance of manned straddle carriers (SC) and automated stacking cranes using simulation models. Petering et al. (2009) showed that appropriate YC dispatching policies have a major influence on a terminal's long-run average quay crane rate using a fully-integrated, discrete-event simulation model. Scheduling algorithms for multiple YCs are also being addressed in the literature (Stahlbock and Voß 2008). For instance, Kim et al. (1999) developed a mixed integer programming model to schedule the load sequence for a single SC in a port container terminal. The objective is to minimize the total travel distance of the SC by optimally determining the SC visit sequence. Lee et al. (2007) considered the scheduling problem of two transtainer systems. The container bay visit sequence and the number of containers picked up at each visit of the two transtainers are determined. A simulated annealing (SA) algorithm is developed to solve the proposed mathematical model. However, they do not generate detailed work schedules for the YCs and individual containers. Kim et al. (2003) sequenced the storage and retrieval jobs for a single YC in an YC block using dynamic programming and heuristics. In addition, Lim et al. (2004) also studied quay crane scheduling with spatial and separation constraints. The problem was modeled as a bipartite graph matching. They showed that a squeaky wheel optimization with a local search approach gives good results within a short time. Zhu and Lim (2006) proved that the crane scheduling problem is NP-complete. They designed a SA framework with a graph-search-based neighborhood search to solve quay crane scheduling problems with noncrossing constraints. Near optimal solutions for instances of various sizes can be obtained. Recently, Cao et al. (2008) formulated a discrete-time integer model for a double-rail-mounted gantry (DRMG) crane scheduling problem which involves different crane interferences. In their paper, two DRMG cranes can share the same handling lane but cannot work simultaneously at the same slot. Defining a one-time period as the time required for an RMG to travel the distance of a single slot, the total time periods involved in their model can be very large in a 2-h time window. A combined scheduling heuristic is designed to solve the problem.

Ng (2005) has proposed a heuristic for constructing detailed work schedules for multiple YCs in a block with intercrane interference, albeit without retrieval moves and minimum crane-distance separation. However, most of the algorithms applied on the YC scheduling problem (Ng 2005; Ng and Mak 2005; Ng and Tsang 2005) only consider storage moves, presumably as it is easier to develop algorithms for storage moves. Recently, Cao et al. (2010) solved an integrated PM and YC model for retrieval jobs using Benders' decomposition but they did not consider YC interference. Wen et al. (2010) developed approaches for scheduling multiple cranes traveling over a shared network of tracks in a shipbuilding environment. Intercrane interference is the major factor affecting makespan and crane utilization. They modeled the problem as a multicommodity flow problem with side constraints and proposed Priority/Ant Colony Optimization (ACO) heuristics to solve the problem. Chen and Lu (2012) addressed the storage location assignment problem for outbound containers. The problem is decomposed into two stages. The yard bays and the amount of locations in each yard bay are first determined by a mixed integer programming model and the exact storage location for each container is obtained with a hybrid sequence stacking algorithm in the second stage. We note that the results from Chen and Lu (2012) may be used as inputs for this chapter.

Models in Chap. 3 and most of the models in the literature (Ng 2005; Petering et al. 2006) which more or less considered YC operational constraints are discrete time. Discrete time models discretize the time horizon into a number of time intervals of uniform duration. Events such as the beginning and end of a task are associated with the boundaries of these time intervals. Only a few papers model the problem as a simplified continuous-time model (Ng and Mak 2005; Ng and Tsang 2005) as shown below:

$$\text{Minimize} \sum_{m \in S_m} \text{Ts}_m - \text{Tgt}_m$$

Subject to:

$$\text{Ts}_m \geq \text{Tgt}_m, \quad \forall m \in S_m$$
$$\text{Ts}_n - \text{Ts}_m \geq d_{mn} + h_m - (1 - X_{mn}) * \text{BM}, \quad \forall m \in S_m, \forall n \in S_m$$
$$X_{mn} + X_{nm} = 1, \quad \forall m \in S_m, \forall n \in S_m$$

where d_{mn} is the time an YC travel from the location of job m to job n and h_m is the YC handling time of job m. The binary variable X_{mn} is 1 if job m precedes job n and 0 otherwise. In the model above, only storage jobs are modeled. As there is no variable to assign different YCs to jobs, only one YC is considered. Besides job sequencing constraints, other operational constraints (e.g., YC interference) are ignored. An YC schedule from this simplified continuous-time model may not be directly applicable to actual YC operations. This chapter develops an efficient continuous-time MILP model and applies appropriate heuristics and algorithm to find near optimal solutions. Detailed CT operating constraints are modeled and the solution quality is greatly increased compared to the discrete-time MILP models.

4.1.3 Contributions

In Chap. 3, a discrete-time model was developed for YC scheduling by taking into account the realistic operational constraints of a large CT such as the intercrane interference, fixed YC separation distances, and simultaneous container storage/retrievals. The solution quality and solution time using a RHA are both better than those found in the literature. Studying the same problem but using an improved model, we extend this work in this chapter with the following fivefold contributions:

(1) **Significantly improved solution quality**.

The continuous-time model significantly improves the solution quality compared to a discrete-time model solving the same dataset. Models in Chap. 3 and most models in the literature (Ng 2005; Petering et al. 2006) which consider YC operational constraints are discrete time. Discrete-time models discretize the time horizon into a number of time intervals of uniform duration. Events such as the beginning and end of a task are associated with the boundaries of these time intervals. As the actual handling times of the container moves are continuous, discrete-time MILP models can only approximate the scheduling problem. One usually needs to use a sufficiently small time interval, for example, the greatest common factor of the processing times to achieve a suitable approximation of the original problem. However, this will significantly increase the number of integer variables associated with each discrete-time interval leading to an intractable model size and much longer computational time. The main limitation of a discrete-time MILP model is its discrete approximation of the continuous-time horizon. Avoiding synchronizing container moves to discrete-time grids, a continuous-time model schedules YC movements to their earliest possible times leading to significantly reduced PM waiting times. However, despite the advantages of a continuous-time model, only a few papers model the YC scheduling problem as a continuous-time model due to the difficulty in problem formulation (Ng and Mak 2005; Ng and Tsang 2005). Also, these continuous-time models consider only storage jobs and only one YC. Besides the job sequencing constraints, other operational constraints (e.g., YC interference) are ignored. An YC schedule from these continuous-time model may not be directly applicable to actual YC operations. Thus, this chapter develops an efficient continuous-time MILP model taking into account realistic CT operating constraints.

(2) **Ability to handle large size problems**.

Without discretizing the time horizon, the number of binary variables is significantly reduced in the continuous-time MILP model developed in this chapter. As a result, the model size is reduced and larger instances with the number of jobs ranging from 45 to 69 in a 2-h time window can be solved quickly in this chapter. This is much larger than the size of the instances handled (20–30 jobs) in Chap. 3. From the YC operational data collected from the CTs, the number of jobs can range from 0 to 60 at each block (Ng 2005). Thus our model is applicable to actual YC scheduling situations.

(3) **Reduced solution time**.

Applying appropriate heuristics and extending the RHA proposed in Chap. 3 to the continuous-time model, the solution time to the same datasets (20–30 jobs) are faster (1.6 s on average in this chapter compared to 4.6 s on average in Chap. 3), mainly due to the reduced model size.

(4) **More realistic operational constraints considered**.

The following constraints are considered for the first time in this chapter as they have been ignored in the discrete-time models (see for example Chap. 3).

- The minimum YC gantry time is defined for consecutive moves at different YCs by taking into account the idle YCs that sit in between two working YCs and four cases of YC order, and slot difference (see Sect. 4.2.2.4).
- When container moves are handled by neighboring YCs, the definition of the corresponding set is updated so that it varies by the order number of the YCs (see Sect. 4.2.2.5). This update takes into account the distance of the idle YCs that sit in between two working YCs.
- For jobs located at the leftmost/rightmost slots, besides the constraints on the first and the last YC (Chap. 3), additional constraints are enforced to prevent certain YCs from handling them (see Sect. 4.2.2.11). Enforcing these implicit constraints originating from the multiple intercrane interference, operational constraints explicitly in the model can reduce the number of binary variables to be determined and hence significantly reduce the solution time.

(5) **Handling last-minute jobs**

Last-minute container arrivals into the CT disrupt the normal CT operations and may impede CT performance. A CT has to handle these jobs and recover to normal CT operations as soon as possible. Two methods in the RHA are discussed and compared in this work to schedule last-minute container arrivals.

This chapter is organized as follows. Section 4.1 provides the introduction and literature review of the crane scheduling problem. The contributions of this chapter are summarized. Section 4.2 presents an efficient continuous-time MILP model. The assumptions applied in the model and computational results for smaller instances are provided and compared with the literature. In Sect. 4.3, heuristics and a RHA are proposed to solve large problems. Section 4.4 discusses approaches for handling last-minute jobs. Section 4.5 presents some discussions and Sect. 4.6 provides a summary of the results and concludes the chapter. Finally, Appendix 1 defines the sets, variables, and parameters and Appendix 2 presents the constraints of the subproblems in RHA.

4.2 Continuous-Time Model (CMIP1)

We now develop a continuous-time MILP model taking into account all YC operating conditions. As the YCs handling and gantrying activities are realistically modeled, CT applicable and optimal schedules are guaranteed. Without defining the discrete-time intervals, the size of continuous-time model is smaller than its discrete-time based counterpart (Chap. 3). Much larger size problems (up to 69 jobs in this work versus at most 32 jobs in Chap. 3) are handled in this work allowing it to cope with realistic YC operating conditions. Our model can deal with either multiple YC or single YC scenarios with minor modifications to the constraints. In the following sections, we present our formulations based on the multiple YC scenario. If changes are needed to suit the single YC scenario, we will point out the differences of the formulations for the single YC scenario for completeness sake.

4.2.1 Objective Function

Appendix 1 contains the notations used for the indices, sets, parameters, and variables in the mathematical formulation whose objective is to minimize a linear combination of the total amount of retrieval earliness, and storage and retrieval delays,

$$\text{Minimize TC} = W_{\text{re}} \sum_{m \in R} \text{RE}_m + W_{\text{rd}} \sum_{m \in R} \text{RL}_m + W_{\text{sd}} \sum_{m \in S} \text{STL}_m \qquad (4.1)$$

Minimizing TCT is used as the objective in most of the literature (Ng 2005). Only for problems with pure storage moves, minimizing TCT is equivalent to minimizing PMWT. If retrieval moves are involved, minimizing TCT will cause retrieval moves to be scheduled as early as possible and the PMWT becomes large. Further, minimizing TCT glosses over the true gap away from the optimal solution, as the deviation from the optimal TCT decreases with the increase in job arrival and job finish times. In Eq. (4.1), increasing the weight for RL, W_{rd}, the model may increase the total RE to reduce the total RL. This may cause the TCT and PMWT of the optimal solution to be larger than the suboptimal solution. In this case, TC as applied in Eq. (4.1) is the best performance measure.

4.2.2 Constraints

We now present the following constraints applicable to YC scheduling. Without any loss of generality, we assume that the slots of a block are ordered from left to right and the moves are ordered according to the slot number of their location, i.e., if $m < n$, then $\text{slot}_m \le \text{slot}_n$. The YCs are also ordered from left to right as they cannot cross gantry each other.

4.2.2.1 A Container Move Should Be Assigned to a YC, i.e.,

$$\sum_{c \in C} W_{mc} = 1, \quad \forall m \in M, \text{NC} > 1 \qquad (4.2)$$

With only one YC, W_{mc} is fixed at 1, effectively deleting this variable. To maintain the consistency of constraints across multi/single YC scenarios, we keep this variable and fix the value of W_{mc} to 1, i.e., $W_{m,c1} = 1, \quad \forall m \in M,\ \text{NC} = 1$.

4.2.2.2 Identifying Overlapping Container Moves *n* Against Move *m*

Different from the discrete-time formulations in which nearby container moves are synchronized to the boundaries of time intervals, container moves can overlap to any extent in a continuous-time formulation. Overlapping container moves (referred to as n, $n \neq m$) against move m are moves whose job finish times, Te_n ($=\mathrm{Ts}_n + \mathrm{HT}$), are greater than the job start time of move m, Ts_m, and their job start time, Ts_n, is smaller than the job finish time of move m, Te_m. Figure 4.1 illustrates the definition of overlapping and nonoverlapping container moves where each box represents a container move with corresponding job start/finish times shown at the left/right side, respectively. Figure 4.1 shows that only moves 3 and 4 are overlapping moves against move m. To identify the overlapping moves, 0–1 variables $A1_{mn}$ and $A2_{mn}$ are defined:

$$A1_{mn} = \begin{cases} 1 & \text{if } \mathrm{Te}_n > \mathrm{Ts}_m \\ 0 & \text{otherwise} \end{cases}$$

That means if move n happens before move m and they are nonoverlapping, then $A1_{mn} = 0$. For example, move 1 in Fig. 4.1. Also,

$$A2_{mn} = \begin{cases} 1 & \text{if } \mathrm{Ts}_n < \mathrm{Te}_m \\ 0 & \text{otherwise} \end{cases}$$

That means if move n happens after move m and the two moves are nonoverlapping, then $A2_{mn} = 0$, for example, move 6 in Fig. 4.1.

For moves whose job finish time equals the job start time of move m, i.e., $\mathrm{Te}_n = \mathrm{Ts}_m$, or moves whose job start time equals the job finish time of move m, i.e.,

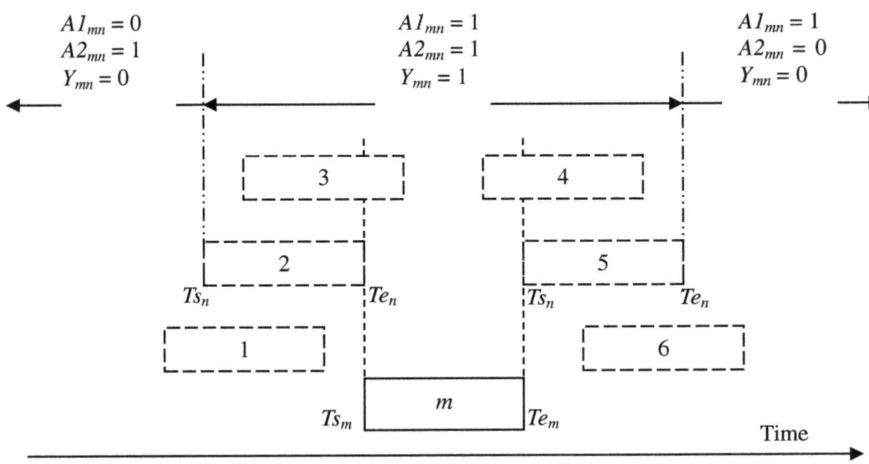

Fig. 4.1 Overlapping and nonoverlapping container moves

$Ts_n = Te_m$, we define these moves as nonoverlapping moves against m. For example, moves 2 and 5 in Fig. 4.1 are defined as nonoverlapping moves against move m. It is clear from Fig. 4.1 that, both $A1_{mn}$ and $A2_{mn}$ of overlapping moves take a value of 1. We can then identify the overlapping moves using $A1_{mn}$ and $A2_{mn}$ and constraints (4.3a)–(4.4b) as follows:

$$\mathrm{Ts}_n + \mathrm{HT} \geq \mathrm{Ts}_m - \mathrm{BM} * (1 - A1_{mn}), \quad \forall m, n \in M,\ m \neq n,\ \mathrm{NC} > 1 \tag{4.3a}$$

$$\mathrm{Ts}_m \geq \mathrm{Ts}_n + \mathrm{HT} - \mathrm{BM} * A1_{mn}, \quad \forall m, n \in M,\ m \neq n,\ \mathrm{NC} > 1 \tag{4.3b}$$

$$\mathrm{Ts}_m + \mathrm{HT} \geq \mathrm{Ts}_n - \mathrm{BM} * (1 - A2_{mn}), \quad \forall m, n \in M,\ m \neq n,\ \mathrm{NC} > 1 \tag{4.4a}$$

$$\mathrm{Ts}_n \geq \mathrm{Ts}_m + \mathrm{HT} - \mathrm{BM} * A2_{mn}, \quad \forall m, n \in M,\ m \neq n,\ \mathrm{NC} > 1 \tag{4.4b}$$

Constraints (4.3a) and (4.3b) define the value of $A1_{mn}$. For move 1 in Fig. 4.1, $Ts_n + HT < Ts_m$, this forces $A1_{mn}$ to take a value of zero, otherwise (4.3a) is violated. Constraint (4.3b) is trivially satisfied when $A1_{mn}$ is zero. For move 3 in Fig. 4.1, the value of $A1_{mn}$ is forced to 1 as $Ts_n + HT > Ts_m$. Similarly, constraints (4.4a) and (4.4b) are used to define the value of $A2_{mn}$. Note that constraints (4.3a)–(4.4b) are only used for scenarios with multiple cranes. For scenarios with a single YC, all moves are handled by one YC and thus there are no overlapping moves (a YC can only handle one container move at a time). Constraints (4.4a)–(4.4b) can be applied to single YC scenarios by eliminating HT and changing the condition $NC > 1$ to $NC = 1$. For single-crane scenarios, $A1_{mn}$ is not needed and we can set its value to 0, i.e., $A1_{mn} = 0,\ \forall m, n \in M,\ m \neq n,\ \mathrm{NC} = 1$.

Figure 4.1 shows that, if move n is used as the reference move, then when $A1_{mn} = 0$, we have $A2_{nm} = 0$; when $A1_{mn} = 1$, we have $A2_{nm} = 1$, etc. In short, $A1_{mn} = A2_{nm}$ and $A2_{nm}$ can replace $A1_{mn}$. Thus, $A1_{mn}$ and constraints (4.3a)–(4.3b) can be eliminated from the model for either multi or single-crane scenarios. Figure 4.1 also shows that, when both $A1_{mn} = 1$ and $A2_{mn} = 1$, moves m and n overlap. Thus, we can use $(A1_{mn} + A2_{mn})$, or equivalently $(A2_{nm} + A2_{mn})$, to identify overlapping moves against move m.

4.2.2.3 Relation of Job Start Times of Consecutive Container Moves at the Same YC

If move n happens after m, and both jobs m and n are handled by the same YC, then constraint (4.5), valid for both multi/single-crane scenarios, holds:

$$\mathrm{Ts}_n \geq \mathrm{Ts}_m + \mathrm{HT} + \mathrm{GH}_{mn} - [3 - W_{mc} - W_{nc} - (1 - A2_{mn})] * \mathrm{BM}, \quad \forall m, n \in M,\ m \neq n,\ \forall c \in C \tag{4.5}$$

If both moves m and n are handled by YC c (i.e., $W_{mc} = W_{nc} = 1$), and move n happens after m (i.e., $A2_{mn} = 0$), then job n should start after the job finish time of move m, Ts_m + HT, plus the YC gantry time from slot_m to slot_n, GH_{mn}. The parameter GH_{mn} is defined as:

$$\mathrm{GH}_{mn} = \begin{cases} |\mathrm{slot}_m - \mathrm{slot}_n| * 0.06 + 0.09 & \text{when } \mathrm{slot}_m \neq \mathrm{slot}_n \\ 0 & \text{when } \mathrm{slot}_m = \mathrm{slot}_n \end{cases}$$

$|\mathrm{slot}_m - \mathrm{slot}_n|$ is the absolute slot number difference. If moves m and n are handled by the same YC consecutively and they are located at the same slot ($\mathrm{slot}_m = \mathrm{slot}_n$), then the YC does not need to move and the gantry time is zero. If the moves are located at different slots, the gantry time (in minutes) of the YC is $|\mathrm{slot}_m - \mathrm{slot}_n| * 0.06 + 0.09$ by taking into account the YC acceleration/deceleration stages.

4.2.2.4 Relation of Job Start Times of Consecutive Container Moves at Different YCs

If move n happens after move m, and moves m and n are handled by different YCs, then constraint (4.6) holds:

$$\mathrm{Ts}_n \geq \mathrm{Ts}_m + \mathrm{HT} + \mathrm{OT}_{mn} - [3 - W_{mc} - W_{nc'} - (1 - A2_{mn})] * \mathrm{BM}, \quad \forall m, n \in M,\ \forall c, c' \in C,\ m \neq n,\ c \neq c',\ \mathrm{NC} > 1 \tag{4.6}$$

Constraint (4.6), valid for multicrane scenarios, states that if moves m and n are handled by different YCs (i.e., $W_{mc} = 1$, $W_{nc'} = 1$, $c \neq c'$), and move n happens after m (i.e., $A2_{mn} = 0$), then job n should start after the job finish time of move m, Ts_m + HT, plus the YC gantry time, OT_{mn}. The definition of OT_{mn} is shown in

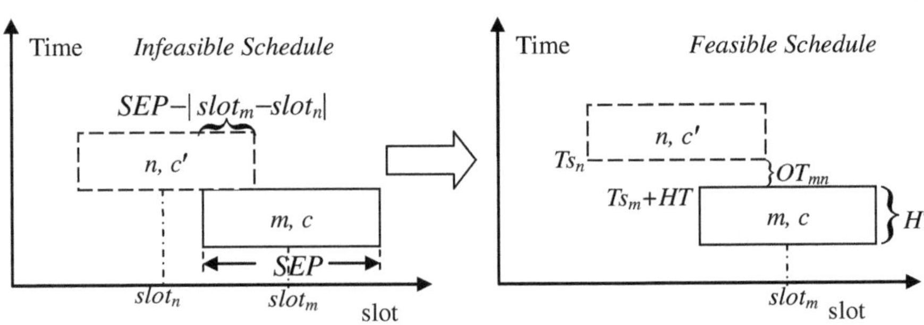

Fig. 4.2 Least gantry time between consecutive moves at different YCs

Fig. 4.2. The height and width of a box in Fig. 4.2 represents the handling time and the influencing slots of a move, respectively. The slot number of a move locates at the center of the box. As when a move is being handled by a YC, a neighboring YC should stay SEP slots away from it, we use SEP to represent the width of a box. On the left of Fig. 4.2, move m is handled by YC c and move n handled by YC c'. Assume $c > c'$. As the slot difference of m and n is less than SEP, YC c' cannot handle move n immediately after YC c finishes handling m. YC c' has to wait for YC c to gantry aside until the slot difference between YC c and slot_n is big enough. If $c - c' > 1$, e.g., move m handled by $c3$ and n handled by $c1$, there is one YC, $c2$ sits between $c1$ and $c3$. We assume that the safety distance between working/idle YCs and working/idle YCs is also SEP. The minimum slot number that YC c has to gantry is $(c - c') * \text{SEP} - |\text{slot}_m - \text{slot}_n|$ and the gantry time is $\text{OT}_{mn} = ((c - c') * \text{SEP} - |\text{slot}_m - \text{slot}_n|) * 0.06 + 0.09$ by taking into account the acceleration/deceleration stages of YC c. The feasible schedule of moves m and n is shown at the right of Fig. 4.2. Note that constraint (4.6) is avoided in the discrete-time model (Chap. 3) by leaving enough travel time at each time interval for YC to gantry to its next job, which could create a longer waiting time for the PM.

The minimum YC gantry time differs by the YC order and slot difference as follows:

$$\text{OT}_{mn} = \begin{cases} (|c - c'| * \text{SEP} - |\text{slot}_m - \text{slot}_n|) * 0.06 + 0.09 \\ \quad \text{when } c > c' \cap \text{slot}_m > \text{slot}_n \text{ or } c < c' \cap \text{slot}_m < \text{slot}_n \\ (|c - c'| * \text{SEP} + |\text{slot}_m - \text{slot}_n|) * 0.06 + 0.09 \\ \quad \text{when } c > c' \cap \text{slot}_m < \text{slot}_n \text{ or } c < c' \cap \text{slot}_m > \text{slot}_n \end{cases} \quad (4.7)$$

One of the four cases in (4.7) is shown in Fig. 4.3 where move m is handled by $c1$ and move n by $c2$ $(c < c')$. Because move n locates at the left of move m $(\text{slot}_m > \text{slot}_n)$, $c1$ will be pushed to the left of move n (m and n still separated by SEP slots) because $c1$ has to stay left of $c2$ all the time. In this case, the total slots $c1$ has to gantry will be $\text{SEP} + |\text{slot}_m - \text{slot}_n|$.

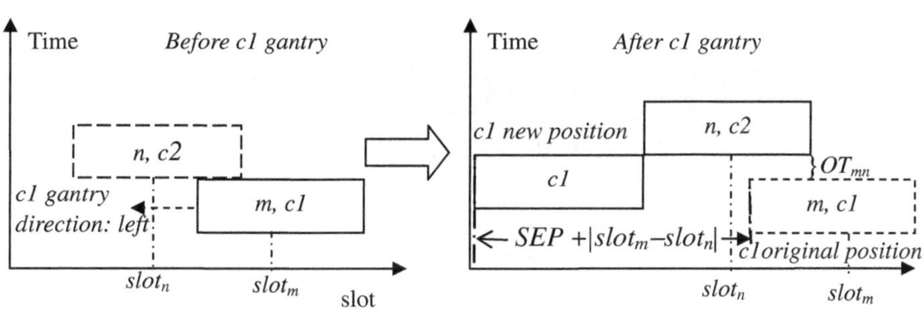

Fig. 4.3 YC minimum gantry slots

4.2.2.5 Moves Handled by Neighboring YCs

For overlapping moves, if they are handled by neighboring YCs, their slot difference should be greater than SEP multiplied by the YC order difference. Constraint (4.8) ensures this:

$$W_{n,c'} \leq 3 - A2_{mn} - A2_{nm} - W_{mc}, \quad \forall c, c' \in C, \forall m, n \in \mathrm{PLJ}_{mn}, \; c' > c, \; c < \mathrm{NC} \tag{4.8}$$

Constraint (4.8), valid for multicrane scenarios, states that if moves m and n are overlapping in time, i.e., $A2_{nm} = A2_{mn} = 1$, and move m is handled by YC c, i.e., $W_{mc} = 1$, then YC c' cannot handle moves which belong to PLJ_{mn}.

Set PLJ_{mn} is defined as $\mathrm{PLJ}_{mn} = \{m, n \in M : (\mathrm{slot}_n - \mathrm{slot}_m) < (c' - c) * \mathrm{SEP} \wedge m \neq n\}$. Hence, the moves that are placed to the left of move m and the moves that are placed to the right of move m but their slot difference against move m is less than $(c' - c)$ * SEP belongs to PLJ_{mn} (Fig. 4.4). Note that PLJ_{mn} varies by the order number of YCs. In Fig. 4.4, YC 2 cannot handle moves that belong to $\mathrm{PLJ1}_{mn}$ when YC 1 is handling move m. If YC 3 is assigned to move n, because there is an idle YC (YC 2) sitting in between YC 1 and 3, $\mathrm{PLJ2}_{mn}$ defined for YC 3 extends SEP slots left compared to $\mathrm{PLJ1}_{mn}$. When moves m and n are far jobs ($|\mathrm{slot}_m - \mathrm{slot}_n| \geq \mathrm{SEP}$) and move n is located to the left of move m handled by YC c (i.e., $W_{mc} = 1$, PLJ_{mn} = true), constraint (4.8) ensures that move n be handled by a YC whose order number is smaller than c when moves m and n overlap in time ($A2_{nm} = A2_{mn} = 1$). If PLJ_{mn} = true and $|\mathrm{slot}_m - \mathrm{slot}_n| < \mathrm{SEP}$, constraint (4.8) ensures that moves m and n do not overlap in time.

In Fig. 4.5, moves $m1$ and $m2$ overlap (at time axis) and they are far jobs because their slot difference is bigger than SEP (at slot axis). As $m2$ locates at the left of $m1$, we have PLJ_{12} = true. If we assume that YC 3 is handling $m1$, constraint (4.8) then ensures that YCs whose order numbers are bigger than 3 (YC 4, YC 5 …) cannot

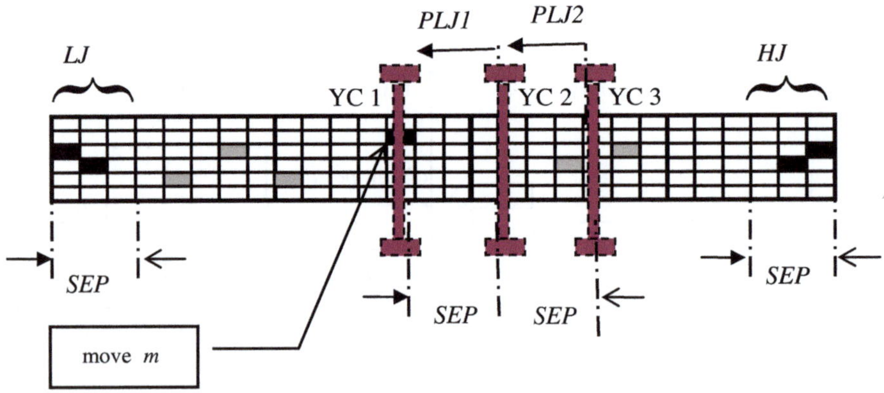

Fig. 4.4 Definitions of PLJ, HJ and LJ

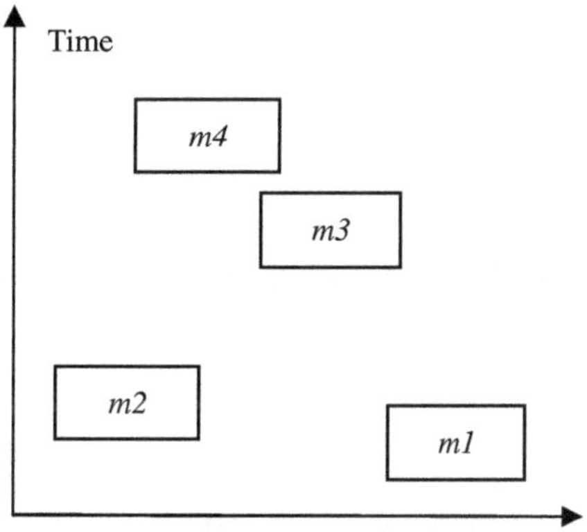

Fig. 4.5 Illustration of constraint (4.8)

handle $m2$., i.e., only YC 1 or YC 2 can handle $m2$. On the contrary, moves $m3$ and $m4$ in Fig. 4.5 are near jobs as $|\text{slot}_3 - \text{slot}_4| < \text{SEP}$. In this case, we have PLJ_{34} = true and PLJ_{43} = true. Applying constraint (4.8) to moves $m3$ and $m4$, respectively, ensures that moves $m3$ and $m4$ cannot overlap in time. Constraint (4.8) is required by SEP apart of YCs and intercrane interference.

4.2.2.6 A YC Can Handle Only One Job at a Time

$$W_{nc} \leq 3 - A2_{mn} - A2_{nm} - W_{mc}, \forall c \in C, \quad \forall m, n \in M, \quad \text{NC} > 1 \tag{4.9}$$

Constraint (4.9) is only valid for multicrane scenarios. It states that if moves m and n are overlapping, i.e., $A2_{nm} = A2_{mn} = 1$, and YC c is handling move m, i.e., $W_{mc} = 1$, then YC c cannot handle move n, i.e., $W_{nc} = 0$.

4.2.2.7 A Storage Move Can Take Place No Earlier Than Its Target Time i.e.,

$$\text{Ts}_m \geq \text{Tgt}_m, \quad \forall m \in S_m$$

4.2.2.8 Computing the Retrieval Earliness, RE_m, of Move m

$$\text{RE}_m \geq \text{Tgt}_m - \text{Ts}_m, \quad \forall m \in R_m \tag{4.10}$$

4.2.2.9 Computing the Retrieval Lateness, RL_m, of Move *m*

$$\mathrm{RL}_m \geq \mathrm{Ts}_m - \mathrm{Tgt}_m, \quad \forall m \in R_m \tag{4.11}$$

4.2.2.10 Computing the Storage Lateness, STL_m, of Move *m*

$$\mathrm{STL}_m \geq \mathrm{Ts}_m - \mathrm{Tgt}_m, \quad \forall m \in S_m \tag{4.12}$$

4.2.2.11 Moves Located at the Firstmost and Rightmost Slots

Moves located at the first SEP slots (Fig. 4.4) can only be assigned to the first YC, due to intercrane interference. Similarly, moves at the last SEP slots can only be assigned to the last YC, namely, $W_{mc} = 1, \forall m \in \mathrm{LJ}_m, c = 1$; $W_{mc} = 1$, $\forall m \in \mathrm{HJ}_m, c = \mathrm{NC}$. Set LJ_m is defined as $\mathrm{LJ}_m = \{m \in M : \ \mathrm{slot}_m \leq \mathrm{SEP}\}$. LJ_m includes moves located at the first SEP slots (see Fig. 4.4). Similarly, HJ_m is defined as $\mathrm{HJ}_m = \{m \in M : \ \mathrm{slot}_m > (\mathrm{NSL} - \mathrm{SEP})\}$ where NSL is the total number of slots. HJ_m includes moves located at the last SEP slots (see Fig. 4.4).

To avoid YC interference, some YCs cannot handle jobs located at the leftmost/rightmost slots. For instance, YC 2 cannot handle jobs in the first SEP slots as YC 1 sits on the left side of YC 2. YC 3 cannot handle jobs in the first 2 * SEP slots as YC 1 and 2 sit on the left side of YC 3. Thus,

$$\begin{aligned} &W_{mc} = 0, \quad c > 1 \cap \mathrm{slot}_m \leq (c - 1) * \mathrm{SEP} \\ &W_{mc} = 0, \quad c < \mathrm{NC} \cap \mathrm{slot}_m > (\mathrm{NSL} - (\mathrm{NC} - c)) * \mathrm{SEP} \end{aligned}$$

Fixing part of W_{mc} as above can reduce the solution time significantly.

4.2.3 Assumptions

The following assumptions are used for the continuous-time models.

- The target times and locations of container moves, found from the QC work list, are assumed known and fixed. Some studies (Chen and Lu 2012) have proposed approaches to optimize the storage locations of container moves. For retrieval jobs, the target time is the latest time by which a retrieval job can be handled and still meets the deadline set by the QCs. For storage jobs, the target time is the earliest time following the release of the job (i.e., the arrival of the corresponding PM) in the yard.

- As a retrieval delay by the YC directly affects the QC schedules and berth operations, and only the storage delay affects the yard operations, the weight for the total retrieval delay, W_{rd}, is set larger. Hence, we set $W_{re} = W_{sd} = 1$ and $W_{rd} = 2$.
- As the job handling time of a YC is usually 2–4 min (Ng and Mak 2005), we set the job handling time of all YCs as 3 min (20 moves/h) (i.e., HT = 3).
- 20 to 60 moves in a 2-h time window are used in the scenarios tested in this chapter. Each yard block contains 40–60 container slots (bays) (i.e., NSL = 40–60).
- The minimum difference in slot numbers allowed for two working/idle YCs at the same time is assumed to be 8 slots (160 ft) (i.e., SEP = 8).
- The YC acceleration/deceleration stages are taken into account when calculating its gantry time. The average YC gantry speed is assumed to be 100 m/min and the acceleration/deceleration is 0.3 ms^{-2}.

4.2.4 Results of CMIP1

CMIP1 includes objective function (4.1), constraints (4.2), (4.4a)–(4.4b), (4.5)–(4.6), and (4.8)–(4.12). To test the performance of CMIP1, we adapted the test scenarios from Chap. 3 which include 30 test cases with 20–32 moves in 40-slot blocks and 30 test cases with 20–32 moves in 60-slot blocks. All test scenarios are randomly generated from a simulator developed for an actual large maritime transshipment terminal (Petering et al. 2009). ILOG CPLEX 11.2 is used to solve all the models developed on a Pentium 1.6 GHz PC. The MILP relative optimality gap of CPLEX is set to 0 (optcr = 0.0). All other settings follow directly from the system default.

Scenario #1, which contains 2 YCs, 40 slots and 20 jobs, is used to illustrate the results. Table 4.1 contains the dataset for scenario #1. The start time of the scheduling

Table 4.1 Data for Scenario #1

Moves	Target time (min)	$slot_m$	Type[a]	Moves	Target time (min)	$slot_m$	Type[a]
m1	49.962	1	R	m11	86.49	24	R
m2	52.962	1	R	m12	89.49	24	R
m3	55.962	1	R	m13	72.336	28	S
m4	34.602	8	S	m14	101.142	30	S
m5	15.806	14	R	m15	38.433	34	R
m6	18.806	14	R	m16	41.433	34	R
m7	21.806	14	R	m17	37.302	35	S
m8	0.992	18	S	m18	109.023	38	R
m9	78.323	22	S	m19	112.023	38	R
m10	83.49	24	R	m20	115.023	38	R

[a]*R* retrieval move; *S* storage move

horizon is zero. Using the formulations described in Sects. 4.2.1 and 4.2.2, and applying the assumptions in Sect. 4.2.3, CMIP1 involves 420 binary variables, 495 single variables, and 3,419 single equations. The model size is much smaller than the discrete-time model (referred to as DMIP which involves 875 binary variables in 47 time intervals, 2,895 variables, and 30,538 single equations) developed in Chap. 3. CMIP1 was solved in 0.3 s with a total PMWT of 4.28 min while DMIP was solved in 1.3 s with a total PMWT of 25.35 min. The PMWT of CMIP1 is 83 % faster than that (25.35 min) obtained from DMIP. The results of scenario #1 from DMIP and CMIP1 are shown in the left and right side of Fig. 4.6, respectively.

Each box in Fig. 4.6 represents a move. The height and width of a box represents the handling time (HT) and the influencing slots (SEP) of a move, respectively. If the actual scheduled start time of a move is different from the original target time, two boxes, a dotted and a solid one, are used to represent the original and the actual schedule, respectively. The handling YC number, job number, and the job type are shown within each solid box. At the right side of each solid box, a letter and a number show whether the actual schedule is delayed ("D") or advanced ("E") and the actual minutes is delayed or advanced. As an example, though the target time of move 8 (the first move at the bottom) is 0.99 min, however, because the beginning of a move in a discrete-time model should be synchronized to the boundaries of time intervals, move 8 is scheduled to start at 3.5 min which is the beginning of the next time interval. It is clear that the above procedure brings a 2.51 min delay to move 8. These delays or earliness resulting from axis synchronization occurs on many other moves in DMIP. The delay on move 8 is completely eliminated in CMIP1 because a move can be handled exactly at its target time if there is no other conflict of operational constraints in a continuous-time model. Scenario #1 is a relatively small example with moves evenly distributed; only 4.28 min of retrieval earliness occurs on move 15 from the results of CMIP1.

After testing the 60 randomly generated scenarios, the mean/maximum/minimum solution times of CMIP1 are 2,016.5/96,111.1/0.29 s, respectively. We note that the solution time of CMIP1 is still unreliable. Table 4.2 summarizes the comparison between the DMIP and CMIP1 using three performance measures. In Table 4.2, TCT is defined as the sum of the completion times of all moves $\left(\mathrm{TCT} = \sum_{m \in M} (\mathrm{Ts}_m + \mathrm{HT})\right)$. The difference between DMIP and CMIP1 is defined as (DMIP-CMIP1)/CMIP1 * 100. Table 4.2 shows that the average difference of PMWT and the objective value between DMIP and CMIP1 is very large. For example, for scenarios with storage and retrieval moves, the average PWMT of DMIP is 399.0 % higher than CMIP1. We note that there is a significant difference between the three performance measures. In this chapter, we will use CMIP1 as the benchmark for our models. The optimal solutions of all test scenarios are obtained by solving CMIP1.

Note that the average TCT gap is 1.0 % by applying the lower bound evaluation procedure proposed by Ng (2005) for pure storage scenarios with 20–32 jobs. Since

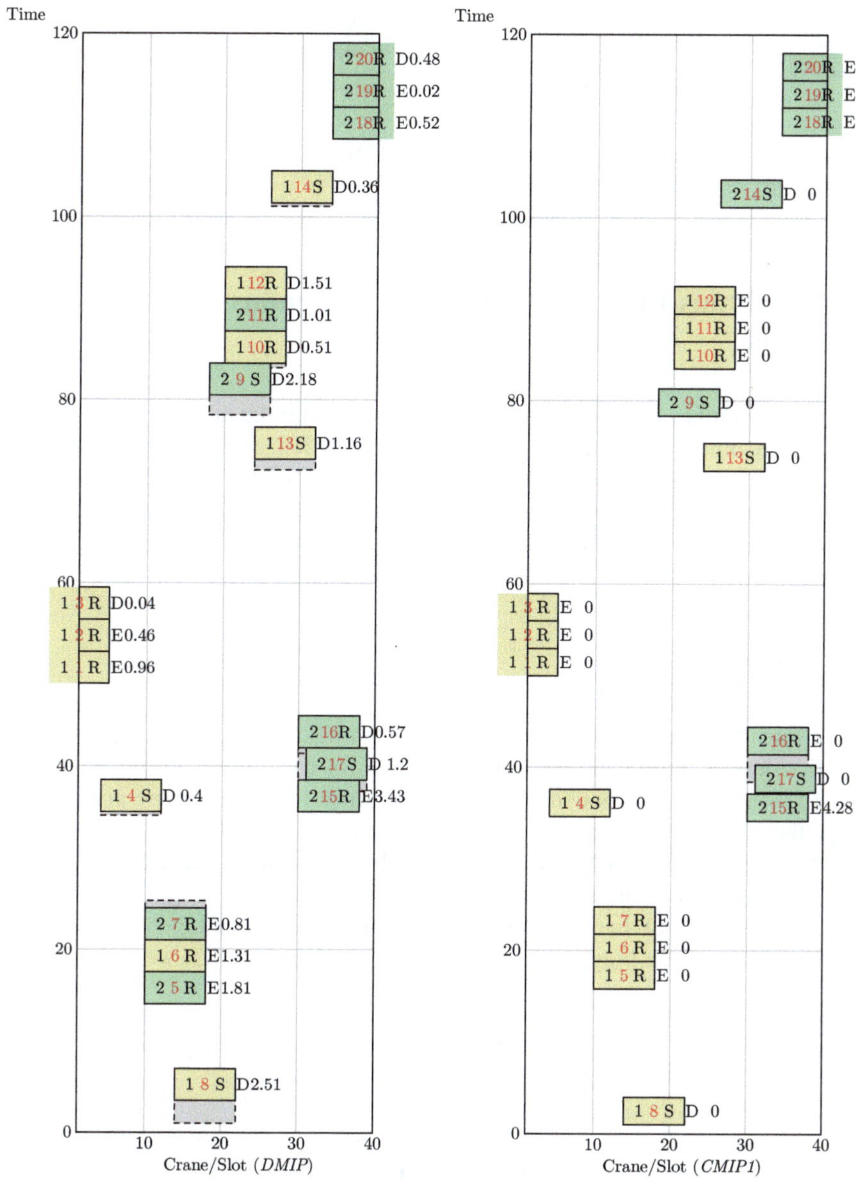

Fig. 4.6 Quality comparison of Scenario #1 between DMIP and CMIP1

we obtain the optimal solutions for all these scenarios by solving CMIP1, the lower bound of TCT gap proposed by Ng (2005) is thus 1.0 % below the optimal solution for scenarios with 20–32 jobs.

Table 4.2 Quality comparison between DMIP and CMIP1

Scenarios	Average gap (%)		
	PMWT[a]	TCT[b]	Objective value
40 slots	519.9	2.5	551.3
60 slots	366.0	3.1	382.3
Pure storage moves	508.9	4.3	508.9
Storage + Retrieval	399.0	1.8	438.7

[a]*PMWT* Total PM Waiting Time; [b]*TCT* Total Completion Time

4.3 Algorithms Integrated with the Continuous-Time Model

Though the number of binary variables is reduced significantly by only defining two bi-index binary variables, W_{mc} and $A2_{mn}$, in CMIP1 compared to defining tri-index binary variables (see Ng 2005), the size of the model still increases significantly with the number of moves. To narrow the search space and reduce the solution time, heuristics and algorithms are developed.

4.3.1 Heuristics

To narrow the search space for CMIP1, a heuristic (Chap. 3) is applied. The heuristics defines a job handling range (T_{lo}, T_{up}) for each move, where T_{lo} and T_{up} are the lower and upper bounds, respectively. Since for both storage and retrieval moves, the greater the difference between the job finish time of a move and its target time, the larger the total PMWT. Thus, the job finish time is scheduled around its corresponding target time as near as possible in an optimal solution. Only moves located in the same job handling range are carefully scheduled together. For moves whose target times are bigger than T_{up}, e.g., move $n1$ in Fig. 4.7, we assume that it

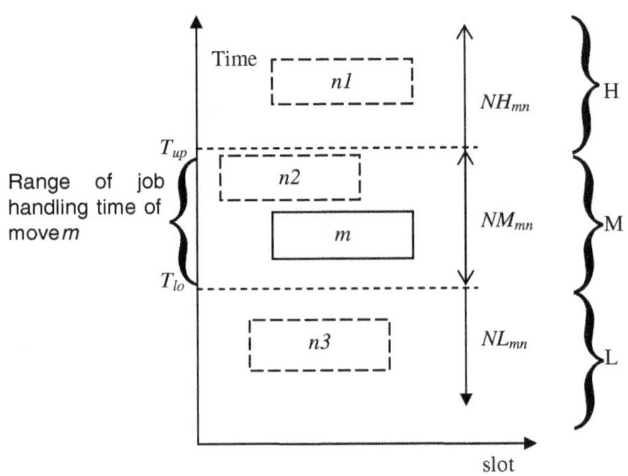

Fig. 4.7 Definition of neighboring moves

must be handled after move m. For moves whose target times are smaller than T_{lo}, e.g., move $n3$ in Fig. 4.7, we assume that it must be handled before move m. Applying this heuristic can significantly reduce the model size because we can fix some binary variables.

To apply the above heuristic to the continuous-time model, we first define sets NM_{mn}, NH_{mn}, and NL_{mn} as follows:

$$\begin{aligned}
\mathrm{NM}_{mn} &= \{m, n \in M \colon |\mathrm{Tgt}_n - \mathrm{Tgt}_m| < \mathrm{JHR} * \mathrm{HT} \wedge m \neq n\} \\
\mathrm{NH}_{mn} &= \{m, n \in M \colon \mathrm{Tgt}_n - \mathrm{Tgt}_m \geq \mathrm{JHR} * \mathrm{HT} \wedge m \neq n\} \\
\mathrm{NL}_{mn} &= \{m, n \in M \colon \mathrm{Tgt}_n - \mathrm{Tgt}_m \leq \mathrm{JHR} * \mathrm{HT} \wedge m \neq n\}
\end{aligned}$$

where JHR is a parameter used to define the job handling range. The value of JHR can be adjusted according to the trade off between the solution quality and time. The solution time decreases with the decrease of JHR, but the solution quality may deteriorate. JHR is set to 4.5 in this chapter. Sets NM_{mn}, NH_{mn}, and NL_{mn} are defined to include moves n whose target times are placed inside/above/below the job handling range of move m, respectively (Fig. 4.7). Some constraints are modified to apply sets NM_{mn}, NH_{mn}, and NL_{mn}. First, constraints (4.4a)–(4.4b) are defined within NM_{mn}:

$$\mathrm{Ts}_m + \mathrm{HT} \geq \mathrm{Ts}_n - \mathrm{BM} * (1 - A2_{mn}), \quad \forall m, n \in \mathrm{NM}_{mn},\ m \neq n,\ \mathrm{NC} > 1 \tag{4ah}$$

$$\mathrm{Ts}_n \geq \mathrm{Ts}_m + \mathrm{HT} - \mathrm{BM} * A2_{mn}, \quad \forall m, n \in \mathrm{NM}_{mn},\ m \neq n,\ \mathrm{NC} > 1 \tag{4bh}$$

Note that the only difference between (4.4a)–(4.4b) and (4ah)–(4bh) is that the condition $\forall m, n \in M$ is replaced by $\forall m, n \in \mathrm{NM}_{mn}$. Since we assume that all moves n in NH_{mn} are handled after move m and all moves n in NL_{mn} are handled before move m, part of $A2_{mn}$ can be fixed i.e.,:

$$A2_{mn} = 0, \quad \forall m, n \in \mathrm{NH}_{mn}; \quad A2_{mn} = 1, \quad \forall m, n \in \mathrm{NL}_{mn}$$

Constraint (4.5) should be replaced by (4.5a)–(4.5c) as follows:

$$\mathrm{Ts}_n \geq \mathrm{Ts}_m + \mathrm{HT} + \mathrm{GH}_{mn} - [3 - W_{mc} - W_{nc} - (1 - A2_{mn})] * \mathrm{BM}, \quad \forall (m, n) \in \mathrm{NM}_{mn},\ \forall c \in C \tag{4.5a}$$

$$\mathrm{Ts}_n \geq \mathrm{Ts}_m + \mathrm{HT} + \mathrm{GH}_{mn}, \ \forall (m, n) \in \mathrm{NH}_{mn}, \quad \forall c \in C \tag{4.5b}$$

$$\mathrm{Ts}_n \leq \mathrm{Ts}_m - \mathrm{HT} - \mathrm{GH}_{mn}, \ \forall (m, n) \in \mathrm{NL}_{mn}, \quad \forall c \in C \tag{4.5c}$$

As constraint (4.5a) works only on moves within NM_{mn}, two more constraints, (4.5b) and (4.5c), are added to ensure that moves within NH_{mn} are handled after move m and moves within NL_{mn} are handled before move m. Similarly, constraint (4.6) should be replaced by (4.6a)–(4.6c):

$$\begin{aligned}\mathrm{Ts}_n \geq \mathrm{Ts}_m + \mathrm{HT} + \mathrm{OT}_{mn} - [3 - W_{mc} - W_{nc'} - (1 - A2_{mn})] * \mathrm{BM},\\ \forall(m,n) \in \mathrm{NM}_{mn}, \forall c, c' \in C,\ m \neq n,\ c \neq c',\ \mathrm{NC} > 1\end{aligned} \tag{4.6a}$$

$$\begin{aligned}\mathrm{Ts}_n \geq \mathrm{Ts}_m + \mathrm{HT} + \mathrm{OT}_{mn}, \forall(m,n) \in \mathrm{NH}_{mn},\\ \forall c, c' \in C,\ m \neq n,\ c \neq c',\ \mathrm{NC} > 1\end{aligned} \tag{4.6b}$$

$$\begin{aligned}\mathrm{Ts}_n \leq \mathrm{Ts}_m - \mathrm{HT} - \mathrm{OT}_{mn},\\ \forall(m,n) \in \mathrm{NL}_{mn}, \forall c, c' \in C,\ m \neq n,\ c \neq c',\ \mathrm{NC} > 1\end{aligned} \tag{4.6c}$$

Again, constraint (4.6) is replaced by three sets of constraints (4.6a)–(4.6c) working on moves within NM_{mn}, NH_{mn}, and NL_{mn}, respectively.

4.3.2 Rolling-Horizon Algorithm (RHA)

As we have observed in Sect. 4.3.1, the job finish time of a move should be scheduled as near as possible around its target time to minimize PMWT. As a result, a later move is likely to be handled after an earlier move, to reduce the gap between the job finish time and the target time. This gives us the opportunity to group moves by their target times. In this chapter, we have extended Chap. 3 RHA and applied RHA to our continuous-time model. In RHA, moves are divided into groups in a rolling-horizon fashion (Fig. 4.8).

Moves with similar target times are grouped since their handling sequence may be interchanged. Moves in a higher group (moves with bigger job target times) will

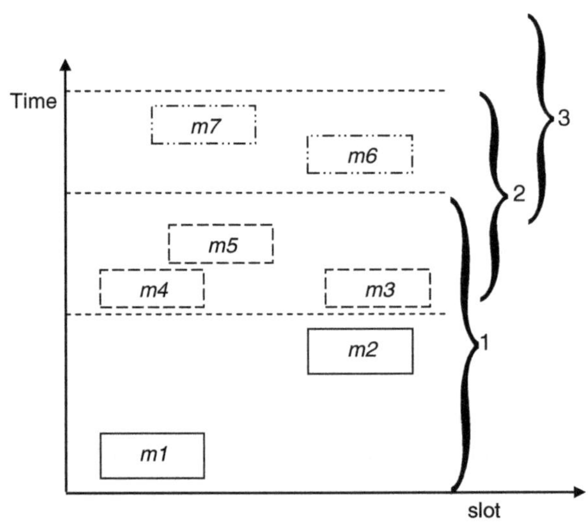

Fig. 4.8 Definition of rolling sub-problems (rename from Fig. 3.5)

be handled later than a lower group (moves with smaller job target times). Thus, the problem can be divided into several subproblems; each subproblem involving moves of one group. As each subproblem involves far fewer moves, it can be solved very quickly. As the handling sequence of moves in a neighboring group which have big target times and moves in a neighboring group which have small target times may also be interchanged, hence all subproblems are generated on a rolling-horizon basis. After a subproblem is solved, two moves with the smallest handling times are fixed. In the following subproblem, two newly added moves together with the nonfixed moves from the previous subproblem are scheduled. If several retrieval moves have very near target times, it is possible to insert one or several of these moves into the empty space among the jobs fixed. Thus, constraints are added to handle the schedules between fixed and free moves. In RHA, set PM_m is defined to include the free moves involved in a subproblem. Then, we redefine sets PLJ_{mn}, HJ_{mn}, LJ_{mn}, NM_{mn}, NH_{mn}, and NL_{mn} under set PM_m. For example, set NM_{mn} is redefined as:

$$\mathrm{NM}_{mn} = \{m, n \in M : |\mathrm{Tgt}_n - \mathrm{Tgt}_m| < \mathrm{JHR} * \mathrm{HT} \wedge m \neq n \wedge \mathrm{PM}_m \wedge \mathrm{PM}_n\}$$

Note that only "$\wedge \mathrm{PM}_m \wedge \mathrm{PM}_n$" is added from the previous definition (Sect. 4.3.1). The MILP model for each subproblem is defined as CMIP2. Appendix 2 lists the other constraints involved in CMIP2. In summary, the MILP model for RHA subproblems (CMIP2) involves two types of constraints:

(i) Constraints for the free moves in a subproblem which include objective function (4.1), (s2), (4ah)–(4bh), (4.5a)–(4.5c), (4.6a)–(4.6c), (4.8), (s9), and (s10)–(s12). All these constraints should be defined under set PM_m.
(ii) Constraints for interactions between all previous fixed moves and current free moves in the subproblem which include (dc4a1), (dc4b1), (dc4a2), (dc4b2), (dc5a), (dc5b), (dc6a), (dc6b), (dc8a), (dc8b), and (dc9).

Figure 4.9 shows the flow diagram of the RHA for the continuous-time model. We set the values of the parameters JNPI (=7) and JHR (=4.5). Set PM_m include moves involved in an iteration. At the start, PM_m includes the first JNPI moves with the smallest target times. CMIP2 is then solved to obtain the schedule for the first JNPI moves. This takes less than 0.1 s to obtain an optimal solution as the size of CMIP2 is very small at each iteration. The job finishing times and handling YCs of the two fixed moves in an iteration are recorded into sets FxTs_m and FxW_{mc}, respectively. The algorithm stops when all moves in SJ_m are scheduled.

4.3.3 Results of the Continuous-Time Model Integrated with Heuristics and RHA

The results of the continuous-time model integrated with the heuristics and RHA (referred to as CRHA) are shown in Tables 4.3 and 4.4.

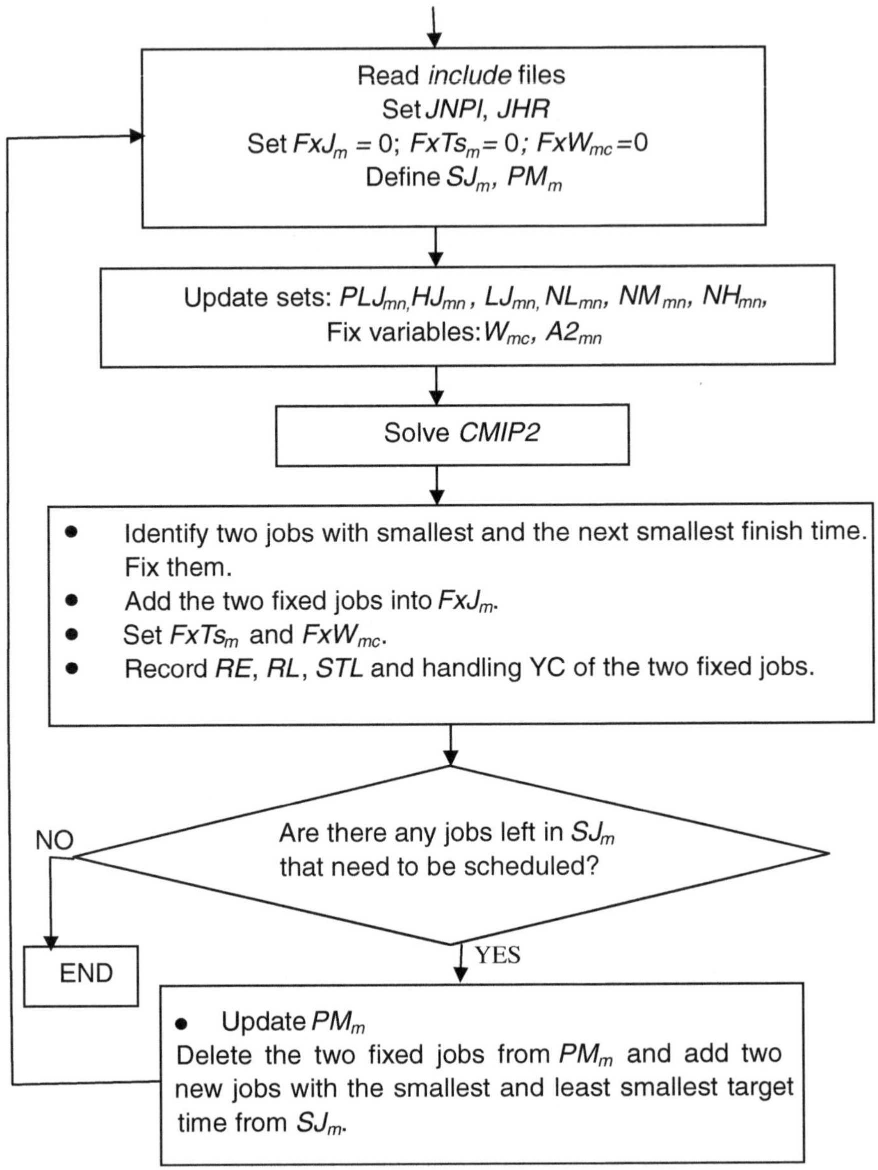

Fig. 4.9 Rolling-horizon algorithm

Table 4.3 shows that CRHA obtains a near optimal solution in less than 2 s on average for scenarios with 20–30 jobs. CRHA is quite robust because the maximum solution time is only 6.0 s. For scenario #1 described in Sect. 4.2.4, CMIP2 involves 29 binary variables, 82 single variables, and 344 single equations at the first iteration. CRHA obtains the optimal solution for scenario #1 in 0.64 s with 8

Table 4.3 Performance (20–30 moves) of CRHA

Scenarios	Solution time (s)			Average gap (%)		
	Mean	Maximum	Minimum	PMWT	TCT	Objective value
40 slots	1.2	3.8	0.6	0.32	0.15	2.21
60 slots	1.6	6.0	0.8	0.82	0.13	3.28
Pure storage	1.1	1.8	0.6	0.02	0.0012	0.02
Storage + Retrieval	1.6	6.0	0.6	0.94	0.24	4.56

Table 4.4 Performance (45–69 moves) of CRHA

Scenarios	Solution time (s)			$(TCT - TCT_{LB}{}^{a})/TCT_{LB} * 100$		
	Mean	Maximum	Minimum	Mean	Maximum	Minimum
40 slots	5.5	15.8	3.0	N/A	N/A	N/A
60 slots	9.1	16.7	2.9	N/A	N/A	N/A
Pure storage moves	6.2	12.6	2.9	3.83	8.18	1.84
Storage + Retrieval	11.4	16.7	5.2	N/A	N/A	N/A

[a]TCT_{LB} lower bound of TCT obtained by applying the approach of Ng (2005)

iterations. The average objective value gap for all scenarios is very small: 0.02–4.56 % (20–30 jobs). A very low average TCT gap (0.0012–0.24 %) is obtained which is much smaller than the gap obtained in the literature (1.2–1.9 % in Ng 2005). Also, in Ng's paper, a discrete-time MILP model was used as the benchmark for his heuristics model. As discrete-time MILP models usually yield solutions whose TCT gap can be 2–4 % higher than the optimal solution (see Table 4.2), the actual TCT gap in his paper is higher than 1.2–1.9 %. CRHA obtains very good solution quality for pure storage scenarios, where the average objective value gap is only 0.02 % and the average TCT gap is 0.0012 %.

To test the performance of CRHA for large instances, we generate 28 test scenarios with moves ranging from 45 to 69. All test scenarios are randomly generated from a simulator developed for an actual large maritime transshipment terminal (Petering et al. 2009). For these large instances with 45–69 moves, the average solution time is about 5–12 s with a maximum solution time of 16.7 s (Table 4.4). CMIP1 yields no optimal solution for these large test scenarios due to the extremely long solution time required. By applying the lower bound evaluation procedure proposed by Ng (2005), the mean TCT gap for pure storage scenarios is 3.83 %. This is smaller than the literature (7.3 % in Ng 2005 and 5.2 % in Chap. 3). Because TCT_{LB} can be 1 % lower than the optimal TCT (see Sect. 4.2.4 for results from 20–32 jobs), the actual average TCT gap for pure storage scenarios is less. Further, a difference in the TCT gap usually corresponds to a much larger difference (about 10 times) in PMWT and the objective gap, the actual PMWT, and the objective gap obtained from this chapter is much smaller than the literature. A key feature of CRHA is that it finds near optimal solutions for scenarios with pure storage moves in 1.1 s (20–30 jobs) and 6.2 s (50–60 jobs) on average.

Although the decrease in size is accompanied by an increase in the number of subproblems to solve, the total solution time still reduces significantly. The results suggest that it is much better to solve a larger number of subproblems than to solve the MILP model singly. As the number of jobs involved at each iteration is set to a small value, the model size of each iteration is very small. As the total problem size increases, the total solution time only increases linearly in n, i.e., $o\left(\frac{N}{\mathrm{JFPI}} * e^{\mathrm{JNPI}}\right)$, With JNPI much smaller than n, the exponential term, e^{JNPI}, is much smaller compared to e^{N}.

4.4 Last-Minute Job Arrivals

Last-minute container arrivals into the CT disrupt the normal CT operations and may impede CT performance. A CT has to handle these jobs and recover to normal CT operations as quickly as possible. Two initial methods in the RHA are discussed in this chapter to schedule last-minute container arrivals as follows.

(1) All moves are fixed except the newly arrived moves. New moves are allocated into the space among all previously fixed moves. As it may yield better solutions by adjusting the handling time of previous scheduled moves, this method may only yield a near optimal solution. As only the free jobs are new moves, the model size is very small and the solution time is thus very short.
(2) Free all moves whose job handling times are larger than the arrival times of the new moves and solve the new problem. This method can find a better solution than (1) albeit a slightly longer solution time. Since we do not need to run iterations whose job handling times are smaller than the new arrival times, the solution time is shorter than solving CMIP2.

In scenario #1, a last-minute container needs to be stored at time 108, slot 34 (referred to as move $m21$). Using approach (1) above, the PMWT yields 14.634 min and the solution time is only 0.03 s. The schedule is shown in the left side of Fig. 4.10. As the handling times of moves $m18$ to $m20$ are fixed, the newly inserted

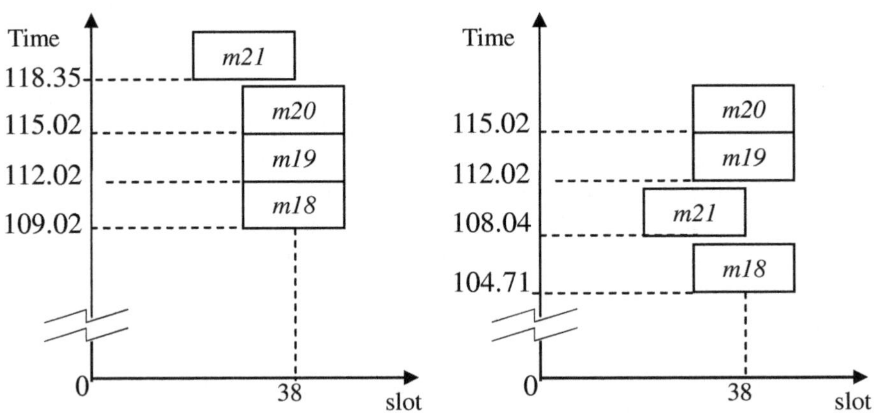

Fig. 4.10 Results of the two approaches handling last minute job insertions

storage move, $m21$, is scheduled to be handled immediately after move $m20$. This introduces a rather large storage delay (10.353 min). By applying approach (2), moves $m18$ to $m20$ are freed and scheduled together with move $m21$. Different from approach (1), $m21$ is inserted between $m18$ and $m19$. The total PMWT is 8.634 min, which is also optimal (confirmed by solving CMIP1). The schedule is shown on the right of Fig. 4.10. The solution time is 0.06 s.

4.5 Discussions

4.5.1 The Best Performance Measure

Minimizing TCT is used as the objective and the deviation away from the optimal TCT is used as the performance measure in most of the literature (Ng 2005). As an objective, only for problems with pure storage moves, the objective of minimizing TCT is equivalent to minimizing PMWT. If retrieval moves are involved, minimizing TCT cannot be used as the objective as this will cause retrieval moves to be scheduled as early as possible and the PMWT becomes large. Further, as a performance measure, TCT glosses over the true gap away from the optimal solution because the deviation from the optimal TCT decreases with the increase in job arrival and job finish times. For example, if the arrival time/optimal job finish time/scheduled job finish time of a storage move is 100/105/110, respectively, and if the job handling time is 3, then the optimal PMWT is 2 (=105-3-100) and the scheduled PMWT is 7 (=110-3-100). While the gap away from the optimal TCT is only 4.7 % (=(110 − 105)/105 * 100 %), the gap away from the optimal PMWT is 250 % (= (7 − 2)/2 * 100 %). The actual average PMWT gap can be much larger than the average TCT gap for pure storage moves (Tables 4.2 and 4.3). In Eq. (4.1), With a TCT gap 1.2–1.9 % of $(TCT - TCT_{IP})/TCT_{IP}$ and a 7.3 % of $(TCT - TCT_{LB})/TCT_{LB}$ achieved in the literature (Ng 2005), where TCT_{IP} is the result of the DMIP, the actual PMWT can be much higher than the optimal PMWT. In an actual CT, PMWT is a greater concern for most operations managers. The deviation from the optimal TCT decreases with the increase in job arrival and job finish times. In the above example, if the job arrival and finish times are assumed to be 200, 205, and 210, respectively, the gap from the optimal TCT decreases to 2.4 %, while the gap from the optimal PMWT remains unchanged. We believe that PMWT is a better performance measure for YC scheduling.

Another concern is the weights for RE, RL, and STL used in the objective function. By increasing the weight for RL, W_{rd}, the model may increase the total RE to reduce the total RL. This may cause the TCT and PMWT of the optimal solution to be larger than the suboptimal solution. In this case, PMWT is not suitable and the best performance measure should be the linear combination of RE, RL, and STL (TC in Eq. (4.1)).

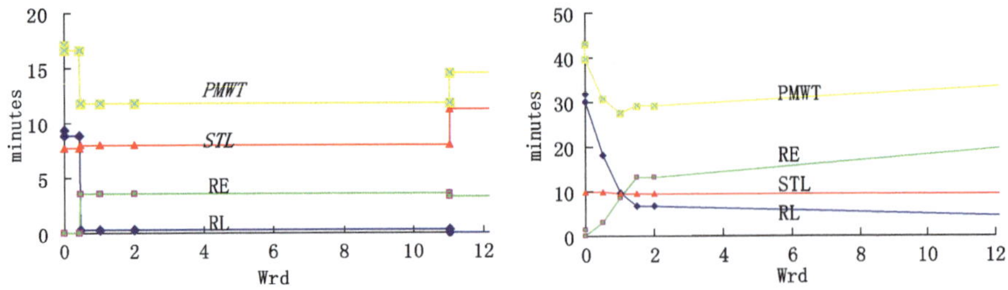

Fig. 4.11 Influence of weights

4.5.2 Influence of Objective Weights

Figure 4.11 shows the influence of the weight used for the total retrieval delay, W_{rd}, on PMWT, RE, RL, and STL for two scenarios obtained using CRHA in this chapter. The two scenarios tested here are the same as the ones used in Sect. 3.6. In general, we find similar trends as in Sect. 3.6 which is obtained using discrete-time models. Because discrete-time models are approximate in nature, the trend in Fig. 4.11 is more accurate.

4.6 Conclusions

Aimed to generate CT applicable and optimal schedules, this chapter develops an efficient continuous-time model and applies a RHA for container YC work schedules. The model size is reduced significantly and the solution time is shortened from days to seconds. The algorithm yields a much higher solution quality in a very short time, saving up to 10 min (for 20–30 moves scenarios) and 50 min (for 50–60 moves scenarios) of the total PMWT compared to other existing heuristic models. A better performance measure is used and discussed in this chapter. The proposed formulations and algorithms may be generally extended to problems with time windows and space constraints such as quay crane scheduling (Lim et al. 2004; Zhu and Lim 2006) or berth allocation (Wang and Lim 2007). Future research will further investigate approaches to last-minute jobs handling and look into developing robust YC scheduling models with uncertain PM arrival times.

Appendix 1: Definition of Sets and Parameters

(a) Indices

c, c' = yard cranes, $c, c' = 1, 2, \ldots, C$
m, n = container moves, $m, n = 1, 2, \ldots, M$

SL = slot number, SL = 1, 2, NSL

(b) Sets

C = number of YCs working in the block.
FxJ_m = jobs that have been fixed after the current iteration and all previous iterations.
HJ_m, LJ_m = includes moves located at the last, first SEP slots in block, respectively.
M = set of container moves to be scheduled.
NL_{mn}, NM_{mn}, NH_{mn} = moves n whose target times are placed below, inside, above job handling range of move m, respectively.
SJ_m = all jobs sorted by their target times.
PLJ_{mn} = moves that locate on the left side of move m and moves that locate at the right side
PM_m = moves involved in a subproblem of RHA.
R_m, S_m = set of yard retrieval, storage moves (the total number of items in R_m and S_m is M).
TSL = number of slots in the block.

(c) Parameters

BM = a large number, $\mathrm{BM} = \mathrm{Max}(\mathrm{Tgt}_m) + 50$
$FxTs_m$ = Job start time (Ts_m) of moves that have been fixed.
FxW_{mc} = W_{mc} of moves that have been fixed.
GH_{mn} = time for a YC to gantry from $slot_m$ to $slot_n$ when consecutive moves m and n are handled by the same YC.
HT = YC handling time of a container move; 3 min is used in this chapter.
JHR = parameter used to defined the job handling range.
JNPI = job number per iteration.
NC = total number of YCs scheduled.
NSL = total number of slots in the block.
OT_{mn} = time for YC to gantry aside when consecutive nearby moves m and n are handled by different YCs.
SEP = minimum difference in slot numbers allowed for two YCs at the same time. SEP is assumed to be 8 slots in this chapter.
$Slot_m$ = Slot number where move m takes place. Without loss of generality, we assume that if $m < n$, then $slot_m \le slot_n$
Tgt_m = target time for move m in the block. For retrieval moves, this is the latest job start time that meets the deadline set by the QCs. For storage moves, this is the earliest job start time following the release of the job.
W_{re}, W_{rd} = weight assigned to total retrieval earliness, total retrieval delay in the objective function, respectively.
W_{sd} = weight assigned to total storage delay in the objective function.

(d) Variables

$A1_{mn}$, $A2_{mn}$ = 0–1 variable to denote if move n happens before, after the start of move m, respectively.
PMWT = total PM waiting time, PMWT = $\sum_{m \in R} \mathrm{RE}_m + \sum_{m \in R} \mathrm{RL}_m + \sum_{m \in S} \mathrm{STL}_m$
RE_m, RL_m = amount of retrieval earliness, retrieval lateness for move m, respectively.
STL_m = amount of storage lateness for move m.
Ts_m = job start time of container move m.
W_{mc} = 0–1, binary variable to denote if container move m is assigned to YC c.
TC = linear combination of the total amount of retrieval earliness, and storage and retrieval delays.
Y_{mn} = 0–1, continuous variable to denote whether moves m and n are overlapping.

Appendix 2: Constraints Used in CMIP2

In CMIP2, the following constraints are modified under set PM_m:
Constraint (4.2) is replaced by (s2):

$$\sum_{c \in C} W_{mc} = 1, \quad \forall m \in \mathrm{PM}_m, \mathrm{NC} > 1 \tag{s2}$$

For single YC scenario, we fix

$$W_{m,c1} = 1, \quad \forall m \in \mathrm{PM}_m, \ \mathrm{NC} = 1$$

Constraint (4.9) is replaced by (s9):

$$\begin{aligned} W_{nc} \leq 3 - A2_{mn} - A2_{nm} - W_{mc}, \\ \forall c \in C, \forall m \in \mathrm{PM}_m, \forall n \in \mathrm{PM}_n, \ \mathrm{NC} > 1 \end{aligned} \tag{s9}$$

Constraints (4.10)–(4.12) are replaced by (s10)–(s12), respectively:

$$\mathrm{RE}_m \geq \mathrm{Tgt}_m - \mathrm{Ts}_m, \forall m \in R_m \cap \mathrm{PM}_m \tag{s10}$$

$$\mathrm{RL}_m \geq \mathrm{Ts}_m - \mathrm{Tgt}_m, \forall m \in R_m \cap \mathrm{PM}_m \tag{s11}$$

$$\mathrm{STL}_m \geq \mathrm{Ts}_m - \mathrm{Tgt}_m, \forall m \in S_m \cap \mathrm{PM}_m \tag{s12}$$

To allow unscheduled jobs to be inserted among the fixed jobs, the following constraints are necessary:

$$\mathrm{Ts}_m + \mathrm{HT} \geq \mathrm{FxTs}_n - \mathrm{BM} * (1 - A2_{mn}), \quad \forall m \in \mathrm{PM}_m, \forall n \in \mathrm{FxJ}_n,\ \mathrm{NC} > 1 \tag{dc4a1}$$

$$\mathrm{FxTs}_n \geq \mathrm{Ts}_m + \mathrm{HT} - \mathrm{BM} * A2_{mn}, \quad \forall m \in \mathrm{PM}_m, \forall n \in \mathrm{FxJ}_n,\ \mathrm{NC} > 1 \tag{dc4b1}$$

$$\mathrm{FxTs}_m + \mathrm{HT} \geq \mathrm{Ts}_n - \mathrm{BM} * (1 - A2_{mn}), \quad \forall n \in \mathrm{PM}_n, \forall m \in \mathrm{FxJ}_m,\ \mathrm{NC} > 1 \tag{dc4a2}$$

$$\mathrm{Ts}_n \geq \mathrm{FxTs}_m + \mathrm{HT} - \mathrm{BM} * A2_{mn}, \quad \forall n \in \mathrm{PM}_n, \forall m \in \mathrm{FxJ}_m,\ \mathrm{NC} > 1 \tag{dc4b2}$$

Constraints (dc4a1), (dc4a2), (dc4b1), and (dc4b2) are the counterparts of constraints (4.4a) and (4.4b), respectively. Note that moves m and n in (dc4a1) and (dc4b1) are defined under sets PM_m and FxJ_n separately and used to determine the value of $A2_{mn}$ only. Thus, (dc4a2) and (dc4b2) are added to determine the value of $A2_{nm}$.

Figure 4.12 shows the definition of sets PM_m and FxJ_n in CRHA and the interactions between them. In Fig. 4.12, moves $n1$ to $n4$ have been solved in previous iterations and added to set FxJ_n. That means, their start times and handling YCs have been fixed. Now, in the current iteration, assume that seven moves are defined in set PM_m and their schedules are to be determined.

Note that besides operation constraints among moves within set PM_m, operating constraints between the previous fixed moves and free moves in PM_m also have to

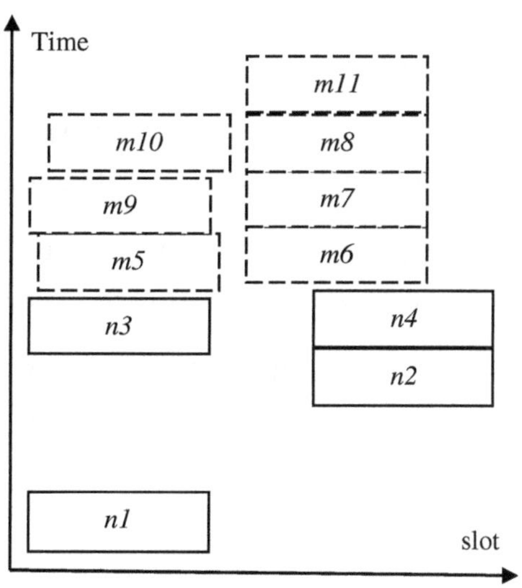

Fig. 4.12 Interactions between free and fixed moves in CRHA

be respected because they are all part of a complete schedule. For example, if YC 2 has been scheduled to handle move *n4* (this decision has been fixed because $n4$ belongs to FxJ_n) and we want YC 2 to handle $m6$ after handling $n4$. The least YC gantry time between $n4$ and $m6$ has to be respected (see Sect. 4.2.2.3 for details). Another important situation is that, if moves in PM_m are very crowded (e.g., moves $m5$ to $m11$ in Fig. 4.12), it might be better to insert some of these moves in the free space of fixed moves, e.g., we could possibly insert move $m5$ after $n1$ and before $n3$ because the YCs are free during that period. If this is the case, then operations constraints between $m5$ and the fixed moves $n1$ and $n3$ should be respected. The following constraints define the operating constraints between free moves and fixed moves. They are counterparts of constraints (4.5)–(4.9) in Sect. 4.2.2, respectively. Constraints (dc5a)–(dc9) differ from (4.5) to (4.9) by defining under sets PM_m and FxJ_n instead. They are added for the completeness of CMIP2.

$$\begin{aligned} Ts_m \geq\ & FxTs_n + HT + GH_{mn} \\ & - [3 - W_{mc} - FxW_{nc} - (1 - A2_{nm})] * BM, \\ & \forall m \in PM_m, \forall n \in FxJ_n, \quad \forall c \in C \end{aligned} \tag{dc5a}$$

$$\begin{aligned} FxTs_n \geq\ & Ts_m + HT + GH_{mn} \\ & - [3 - W_{mc} - FxW_{nc} - (1 - A2_{mn})] * BM, \\ & \forall m \in PM_m, \forall n \in FxJ_n, \quad \forall c \in C \end{aligned} \tag{dc5b}$$

$$\begin{aligned} FxTs_n \geq\ & Ts_m + HT + OT_{mn} \\ & - [3 - W_{mc} - FxW_{nc'} - (1 - A2_{mn})] * BM, \\ & \forall c, c' \in C, c \neq c', NC > 1, m \neq n, \quad \forall m \in PM_m, \forall n \in FxJ_n \end{aligned} \tag{dc6a}$$

$$\begin{aligned} Ts_n \geq\ & FxTs_m + HT + OT_{mn} \\ & - [3 - FxW_{mc} - W_{nc'} - (1 - A2_{mn})] * BM, \\ & \forall c, c' \in C, c \neq c', NC > 1, m \neq n, \quad \forall n \in PM_n, \forall m \in FxJ_m \end{aligned} \tag{dc6b}$$

$$\begin{aligned} & W_{n,c'} \leq 3 - A2_{nm} - A2_{mn} - FxW_{mc}, \quad \forall c, c' \in C,\ c' > c,\ c < NC, \\ & (slot_n - slot_m) < (c' - c) * SEP,\ m \neq n,\ \forall n \in PM_n, \forall m \in FxJ_m, \end{aligned} \tag{dc8a}$$

$$\begin{aligned} & FxW_{n,c'} \leq 3 - A2_{nm} - A2_{mn} - W_{mc}, \quad \forall c, c' \in C, c' > c,\ c < NC \\ & (slot_n - slot_m) < (c' - c) * SEP,\ m \neq n, \quad \forall m \in PM_m, \forall n \in FxJ_n, \end{aligned} \tag{dc8b}$$

$$\begin{aligned} & W_{nc} \leq 3 - A2_{mn} - A2_{nm} - FxW_{mc}, \\ & \forall c \in C, \forall n \in PM_n, \forall m \in FxJ_m,\ NC > 1 \end{aligned} \tag{dc9}$$

References

Cao, J. X., Lee, D.-H., Chen, J. H. and Shi, Q. (2010) The integrated yard truck and yard crane scheduling problem: Benders' decomposition-based methods. *Transportation Research Part E: Logistics and Transportation Review.* 46(3):344–353.

Cao, Z., Lee, D.-H., Meng, Q. (2008) Deployment strategies of double-rail-mounted gantry crane systems for loading outbound containers in container terminals. *International Journal of Production Economics.* 115:221–228.

Chen, L., Lu Z. (2012) The storage location assignment problem for outbound containers in a maritime terminal. *International Journal of Production Economics.* 135:73–80.

Eyefortransport (2008) U-Freight comments on intra-Asia trade report, http://www.eyefortransport.com/content/u-freight-comments-intra-asia-trade-report. accessed Feb. 14 2015.

Froyland, G., Koch, T., Megow, N., Duane, E., Wren, H. (2008) Optimizing the landside operation of a container terminal. *OR Spectrum.* 30: 53–75.

Kim, K.H., Lee, K.M., Hwang, H. (2003) Sequencing delivery and receiving operations for yard cranes in port container terminals. *International Journal of Production Economics.* 84:283–292.

Kim, K. Y., Kim, K. H. (1999) A routing algorithm for a single straddle carrier to load export containers onto a containership. *International Journal of Production Economics.* 59:425–433.

Lee, D.-H., Cao, Z., Meng, Q. (2007) Scheduling of two-transtainer systems for loading outbound containers in port container terminals with simulated annealing algorithm. *International Journal of Production Economics.* 107:115–124.

Lim A., Rodrigues B., Xiao F., Zhu Y. (2004) Crane scheduling with spatial constraints. *Naval Research Logistics.* 51:386–406.

Lim A., Rodrigues B., Xu Z. (2007) A *m*-parallel crane scheduling problem with a non-crossing constraint. *Naval Research Logistics.* 54:115–127.

Ng, W. C., (2005) Crane scheduling in container yards with inter-crane interference. *European Journal of Operational Research.* 164:64–78.

Ng, W. C., Mak, K. L. (2005) Yard crane scheduling in port container terminals. *Applied Mathematical Modelling.* 29:263–275.

Ng, W. C., Tsang, W. S. (2005) Scheduling yard crane in a port container terminal using genetic algorithm. *The First International Conference on Transportation Logistics (T-Log 2005), 27–29 July, Singapore.*

Petering, M. E. H., Wu, Y., Li, W., M. Goh, K. G. Murty, R. de Souza (2006) Simulation analysis of yard crane routing systems at a marine container transshipment terminal. International Congress on Logistics and SCM Systems, Kaohsiung, Taiwan, May 1–2.

Petering, M. E. H., Wu, Y., Li, W., M. Goh, R. de Souza (2009) Development and simulation analysis of real-time yard crane control systems for seaport container transshipment terminals. OR Spectrum. 31(4):801–836.

Stahlbock, R., Voß, S. (2008) Operations research at container terminals: a literature update. *OR Spectrum.* 30:1–52.

Vis, I.F.A., (2006) A comparative analysis of storage and retrieval equipment at a container terminal. *International Journal of Production Economics.* 103(2), 680–693.

Wang, F., Lim, A. (2007) A stochastic beam search for the berth allocation problem. *Decision Support Systems.* 42:2186–2196.

Wen, C., Eksioglu, S. D., Greenwood, A., Zhang, S. (2010) Crane scheduling in a shipbuilding environment. *International Journal of Production Economics.* 124:40–50.

Zhu Y., Lim A. (2006) Crane scheduling with non-crossing constraint. *Journal of the Operational Research Society.* 57(12): 1464–1471.

Chapter 5
Heuristic Approach

The yard crane scheduling problem has been proved to be NP-hard (Bish et al. 2001). As discussed in the earlier chapters in this book, mathematical models can be developed to solve the problem. However, in practice, heuristics are often proposed to solve this optimization problem due to the size and complexity of real-world problems (Sharif and Huynh 2012). This chapter attempts to discuss two underlying concepts for heuristics development for the problem of multiple yard crane scheduling: concurrent and divide and conquer. We then showcase one approach that is based on clustering and reassigning concepts, which applies the divide and conquer approach.

5.1 Concurrent Scheduling

For the multiple yard crane scheduling problem, there are usually many jobs that need to be handled within a relatively short time horizon. The concurrent scheduling approach tries to schedule jobs for all cranes at the same time. These cranes need to maintain the safety distance all the time while considering the different requirements for retrieval and storages jobs, respectively.

Figure 5.1 schematically demonstrates the concurrent scheduling approach. Imagine that all the jobs are placed on a two-dimensional canvas, where the width represents the slots while the height indicates the planning horizon. As displayed in Fig. 5.1, all jobs are plotted on the canvas, together with their job number and the type of jobs: 'R' for retrieval and 'S' for storage. The width of each gray rectangle is 8 slots, which is assumed to be the safety separation requirement. Therefore, at any time, as long as these gray rectangles are not overlapping with each other, they can be handled concurrently. For example, for a two-crane scheduling problem, jobs 5 and 11 could be scheduled concurrently; while job 13 has to wait until job 11 is finished. In this case, job 13 could be handled at the same time with job 24. Jobs can be pushed back to allow the construction of a feasible schedule first.

W. Li et al., *Planning and Scheduling for Maritime Container Yards*,
DOI 10.1007/978-3-319-17025-1_5

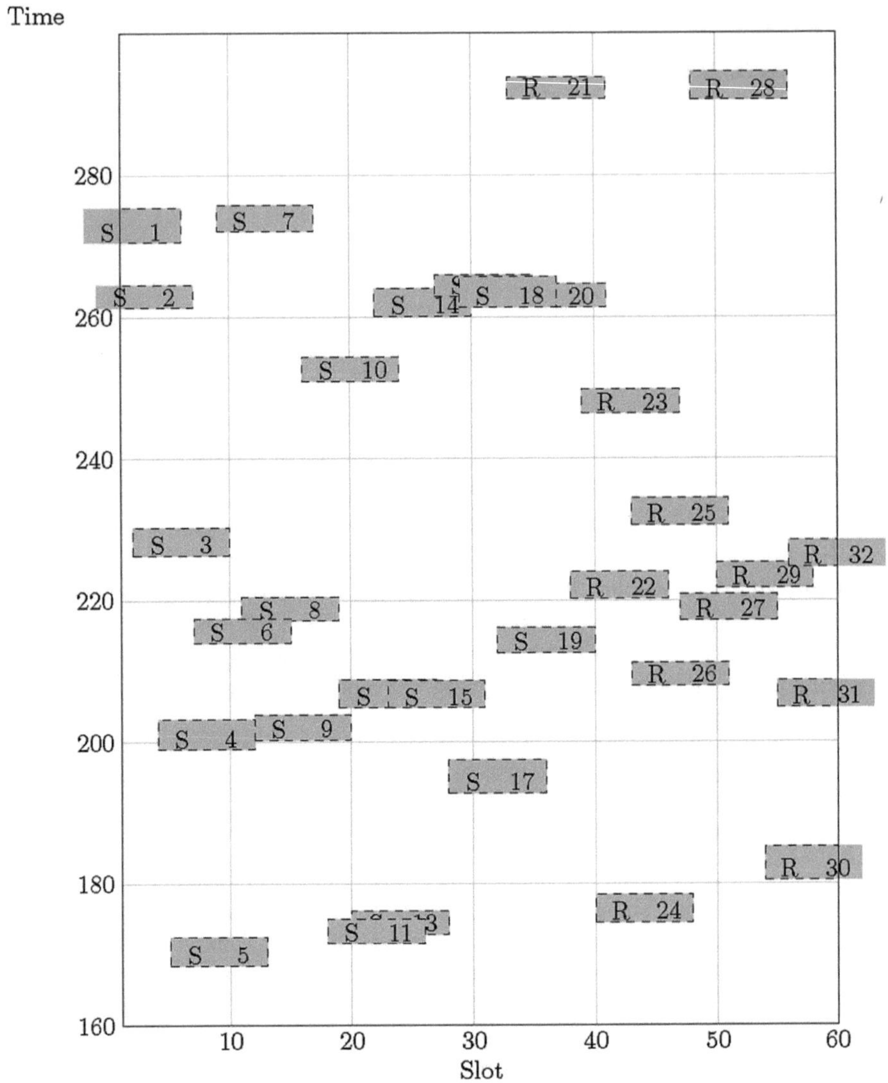

Fig. 5.1 Concurrent scheduling

Essentially, this approach is similar to a rolling horizon approach where the rolling horizon is very small.

Once feasible schedules are obtained for all yard cranes, fine-tuning operations to the schedules can be applied. These operations may include switching the order of jobs for a particular yard crane schedule, exchanging jobs between two yard cranes, and among many others. The ultimate goal is to reduce the retrieval earliness and retrieval and storage lateness to ensure the minimal length of makespan for all yard cranes.

5.2 Divide and Conquer

While the concurrent scheduling approach deals with all cranes at the same time, the divide and conquer approach deals with each yard crane individually, as the name suggests. It includes three major steps: job allocation to different yard cranes, individual schedule generation, and conflict resolving once all schedules are put together.

In the divide and conquer approach, the jobs need to be allocated to different yard cranes first. Some straightforward approaches can be used, such as dividing the jobs evenly to different yard cranes. More sophisticated approaches may consider the 'density' of jobs, i.e., how close the jobs are from each other. For example, Fig. 5.2 illustrates one job allocation based on small clusters, which are formed based on the

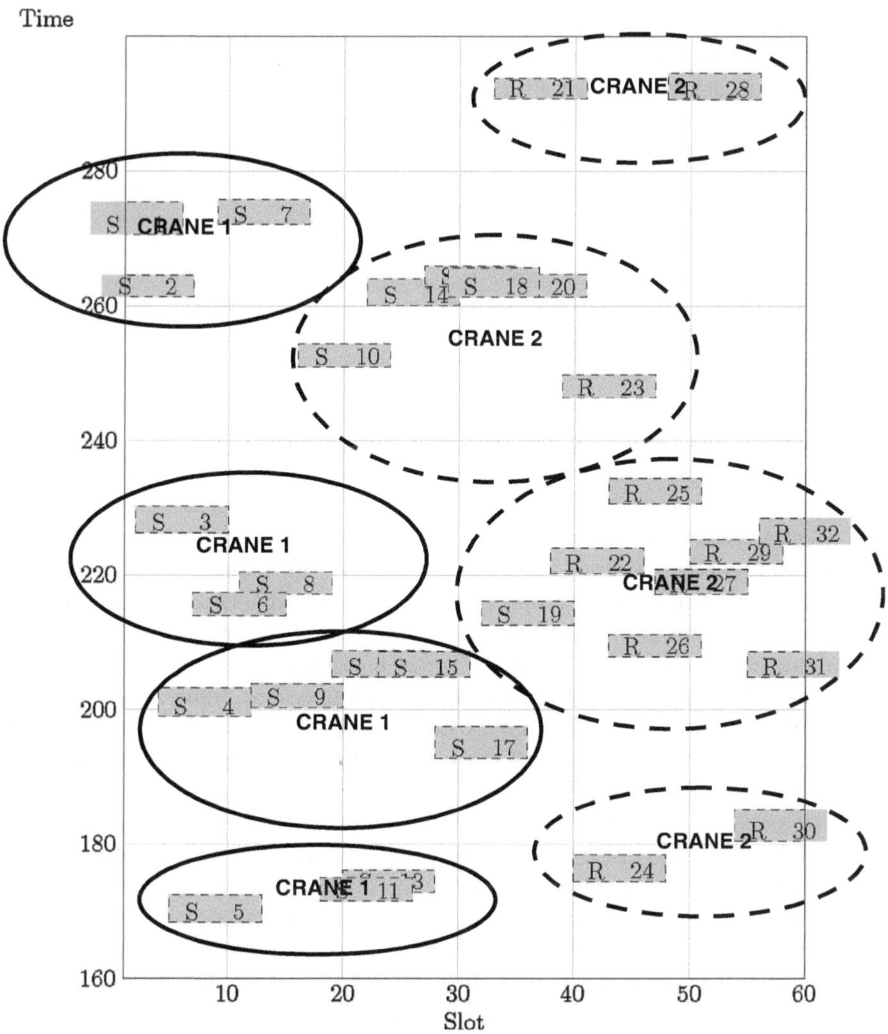

Fig. 5.2 Dividing the jobs to yard cranes based on small clusters

locations of jobs and the distances between jobs. Of course, this is not the final job allocation to all yard cranes and adjustments can be made accordingly, if necessary.

Once the job allocation is completed, each yard crane is ready to generate its own schedule. The scheduling for single yard crane eliminates the problem of yard crane interference and therefore, significantly reduces the complexity of the problem. Each yard crane is given a list of jobs and these jobs need to be arranged in order to minimize the retrieval earliness and retrieval and storage lateness, i.e., incur the minimum disruption to the schedules of other pieces of equipment that work with these yard cranes.

When all the individual yard cranes schedules are combined, the overall feasibility of these schedules should be checked, i.e., ensure there are no inter-crane interferences. If, for any reason, there is interference, then necessary remedies need to be employed and usually compromises will have to be accepted in order to make the overall schedules feasible.

These remedies may include switching the orders of jobs, postponing certain jobs in one particular schedule, or reallocating conflicting jobs to other adjacent yard cranes to achieve an overall feasible schedule for all yard cranes. While it is relatively easy to generate individual schedules for each yard crane, putting all individual schedules together and make them an overall feasible schedule might not be that easy, and generally speaking, the more yard cranes involved, the more difficult to reach the ultimate overall feasible schedule.

5.3 Clustering-Reassigning Heuristic

The clustering-reassigning heuristic described here is based on Wu et al. (2006). It adopts the advantages of both the concurrent scheduling and the divide and conquer approaches. The algorithm models the actual crane travel time realistically and considers the inter-crane interference while generating the yard crane schedules. It first assigns jobs to each yard crane based on the concept of clustering and then addresses the potential conflicts caused between any pair of yard cranes. We start with the introduction of some of the concepts that will be used for the heuristic construction.

The first concept is the *operational continuity requirement*, which controls job starting times of consecutive container moves handled by the same yard crane. For two consecutive moves j and k handled by the same yard crane, without loss of generality, we can assume that move k happens after move j. Then job k can only start after the job finishing time of move j, plus the yard crane gantry time, which allows the yard crane to travel away.

Earliest possible start time refers to the earliest time any job can actually be handled. In the context of retrieval operations, this normally refers to the time ‘0’, i.e., the start time of the planning horizon; for storage jobs, it is the time when the container actually arrives at the slot where it will be stored.

Conflict-prone jobs are defined as these container moves which might cause schedule conflict between any two adjacent yard cranes (i.e. clusters). For two clusters of jobs, which will be handled by two adjacent yard cranes, the maximum slot number in the left cluster and the minimum slot number in right cluster can be identified, respectively. The possible interference is considered by adding the safety distance requirement in terms of number of slots to the maximum slot number in the left cluster, and subtracting the number of slots from the minimum slot number in the right cluster, from which we can get two numbers. Those jobs with slot numbers in between these two numbers are defined as conflict-prone jobs since they will be likely to conflict with each other with the safety distance requirement.

Storage lateness is the amount of time (delay) between the actual handling time and the targeted handling time for a storage job. Since storage job can only be handled after its target time, the minimum storage lateness is 0.

Retrieval earliness is the amount of time ahead of a retrieval job's targeted handling time. Retrieval earliness ensures the retrieval jobs will always be able to meet the quay cranes' schedules by arranging prime mover to arrive at the quay side earlier. However, the amount of retrieval earliness should be kept as small as possible to minimize the waiting time of prime mover at the quay side.

Retrieval lateness is the amount of time (delay) between the actual handling time and the targeted handling time for a retrieval job. Retrieval lateness will directly affect the quay cranes' schedules and therefore should be avoided whenever possible.

5.3.1 The Algorithm

As we know, the inter-crane interference is caused by the safety distance requirement between any pair of adjacent yard cranes. Furthermore, the order of cranes along the yard block is also fixed since no crane can cross the other ones without leaving its block. It should be noted that there are yard cranes that can crossover each other, such as the double-rail-mounted gantry (DRMG) crane system, where the heights of cranes are different to allow the crossover to happen. Scheduling of such equipment, such as Cao et al. (2008) and Dorndorf and Schneider (2010), takes the specific problem characteristics into consideration. The yard cranes under investigation in this chapter, however, are different from those of the DRMGs. Therefore, the service range of one particular crane is constrained by its neighboring cranes, which leads to the intuitive idea that each crane handles a relatively concentrated 'area' (in terms of slot location) of containers.

The clustering-reassigning approach follows this intuition. Algorithm 5.1 presents a brief description of the approach. The algorithm first clusters the job set in a particular yard block into several subsets based on the locations (slots) of the jobs, where the number of subsets is equal to the number of cranes to be scheduled. To divide the jobs into clusters, the *K*-means clustering algorithm (Tou and Gonzalez 1974) is adopted as we already know the number of clusters (i.e., cranes) to be constructed.

Subsequently, each cluster is assigned to a crane according to their relative sequences (locations) along the yard block and schedules are generated within each cluster. In this way, the cranes are guaranteed not to crossover each other when operating. However, due to the requirement of keeping an operational safety distance between every two adjacent cranes, such schedules may not be feasible even when the clusters are not overlapping. We use a novel approach to deal with this problem if the initial schedules are interfering with each other. The jobs that can cause conflict for the yard cranes are first put aside and inserted into a temporary list L_t; the remaining jobs are rescheduled within each cluster, and the resulting schedules are always feasible since no conflict can happen after the conflict-prone jobs are removed. The jobs in the list L_t are then inserted back into the schedules individually by adjusting the operations in the schedule to make room for the newly inserted job to ensure that every newly inserted one does not violate the feasibility of all existing schedules. The possible start times of jobs remaining in the list L_t are subsequently updated according to the newly inserted job. This operation is repeated until the list L_t is depleted. The basic idea of this conflict resolving mechanism is that those jobs that are not in the list L_t are relatively free to be moved within the schedule since they do not affect the feasibility of all schedules, while those in the list L_t are relatively inflexible because they are conflict-prone.

Algorithm 5.1 Clustering-reassigning approach for scheduling multiple YCs

```
Data elements: yard crane number C, temporary list Lt;
1:      clustering the jobs into C clusters;
2:      assign each crane with one cluster;
3:      generate initial schedules for each crane;
4:      if the initial schedules are feasible then
5:          go to step 23;
6:      else
7:          for each pair of adjacent clusters do
8:              identify and take out jobs that are conflict-prone and insert them into a
                temporary list Lt;
9:          end for
10:
11:         for each crane schedule do
12:             re-generate schedule for each cluster;
13:         end for
14:
15:         while Lt is not empty do
16:             take the first job in the list Lt;
17:             insert the job back to its original schedule;
18:             make adjustment so that the newly inserted job can be scheduled;
19:             update the possible start times of jobs in Lt;
20:         end while
21:     end if
22:
23:     for each job meeting reassignment criteria do
24:         reassign it to its neighboring crane job list and generate new schedules;
25:         accept the new schedules if better performance is achieved;
26:     end for
```

In the reassigning part of the algorithm, jobs are bounced around between schedules of neighboring cranes to see if better overall schedules can be achieved. Of course, not all jobs can be reassigned to other cranes. For example, a leftmost job in a cluster would be difficult to be reassigned to a right neighboring crane. Therefore, certain criteria should be applied to eliminate unnecessary computational burden. One might only consider that the left half of the jobs in a cluster should be tried for reassignment to a left neighboring crane and the right half for a right neighboring crane. Of course, other job reassigning criteria could be used subject to user specification. The new schedules are accepted if they generate better results; otherwise, the original schedules are restored.

It should be mentioned that although the yard crane travel time can be modeled realistically, the handling time of each job cannot be known *a priori*. Therefore, the handling time is assumed to be a rather conservative constant amount of time for each job when scheduling. Figure 5.3 depicts the scheduling results of the clustering-reassigning approach for a rather congested job set with two yard cranes in a 40-slot yard block. In the figure, each job is represented as a box where the width of the box is equal to 8 slots and the height of the box is the duration required to handle the move plus the gantry time required for the next job. The actual move is located at the central slot and the box indicates that the crane is "occupying" these slots when handling. Obviously, the schedules will be feasible when there is no intersection among boxes as the interference has been incorporated into the width of the boxes. The gray areas with dashed lines around them are the original operations targeted to be carried out; colored boxes with solid lines around them are the generated schedules, where each color represents one crane's schedule, the numbers "0" and "1" indicate "crane 0" and "crane 1", the letter "S" indicates the operation is a storage one; the blank area in each solid box (if any) is the gantry time that a YC requires to travel to the next job.

The complexity of the clustering-reassigning algorithm is in polynomial order of n, where n is the number of jobs to be scheduled. This gives the algorithm a great advantage to provide online scheduling support to the container terminal operations management. The complexity of clustering of jobs using the K-means approach is in $O(n)$, while the complexity of generating individual schedules is in $O(n^2)$ and the conflict resolving part again has the complexity of $O(n^2)$. Therefore, the whole clustering part of the algorithm has the complexity of $O(n^2) + N_C O(n^2) + O(n) \sim O(n^2)$, where N_C is the number of cranes to be scheduled. In the reassigning part, (at most) each job has the possibility to be reassigned to its left or right neighboring cranes and the whole schedules are regenerated; therefore, the complexity of the clustering-reassigning algorithm would be $O(n^2)n \sim O(n^3)$. As explained, not every job is tried to be reassigned; hence, the complexity can be further written as $o(n^3)$.

5.3.2 Comparison with the Continuous MILP Model

In order to compare the performance of the proposed heuristic algorithm, we conduct numerical experiments on 90 test cases and compare the results with the

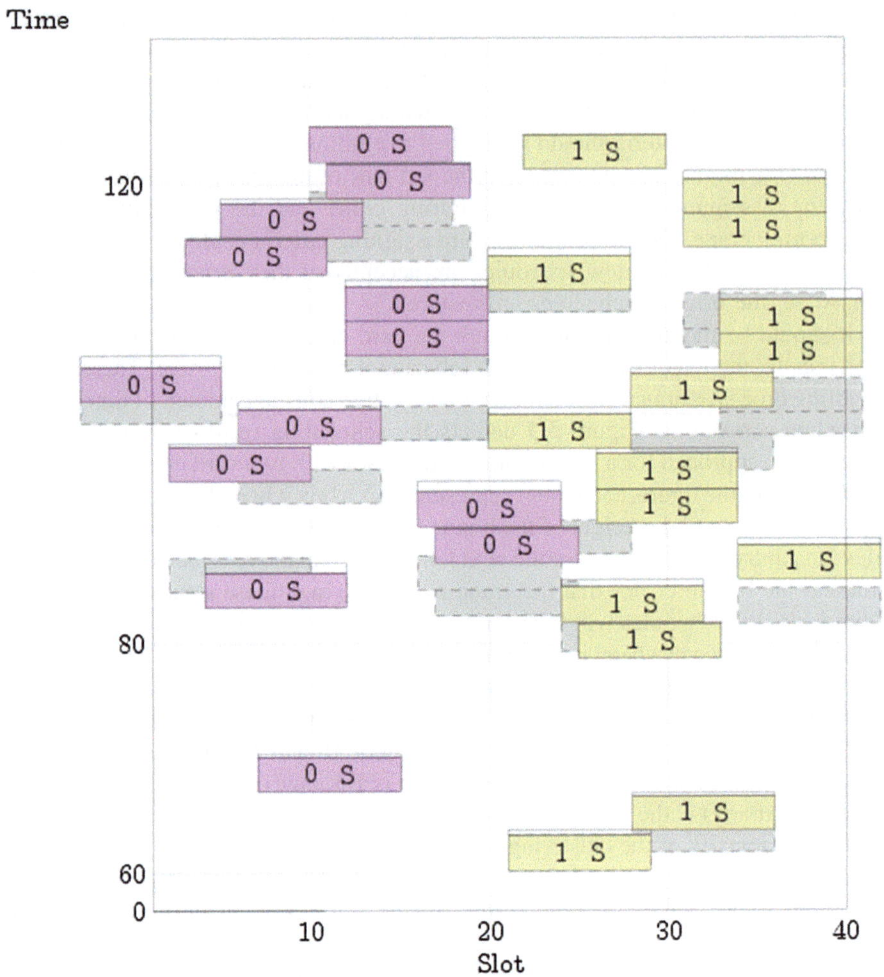

Fig. 5.3 Scheduling results for a job set with two yard cranes

continuous time model developed in Chap. 4. Different yard operation scenarios are considered: pure storage, pure retrieval, and mixed storage/retrieval jobs. Both 40-slot and 60-slot yards are examined. The workload of different cases is also controlled by specifying the number of jobs to be handled. Table 5.1 lists a summary of all the test cases and then classifies these test cases in terms of pure storage, pure retrieval, and mixed storage and retrieval operations.

The mixed integer linear programing (MILP) model developed in Chap. 4 (Li et al. 2012) is used as the benchmark and solved with GAMS/CPLEX 11; the clustering-reassigning approach is coded in C++. Both approaches are tested on Windows XP with a 2.2 GHz Pentium processor and 2 GB RAM. Table 5.2 displays the performance of the clustering-reassigning approach compared with the

Table 5.1 Summary and classification of test cases

Number of cases	Block dimension	Number of jobs	Workload
30	40-slot	19–33	Light
14	40-slot	41–69	Heavy
32	60-slot	19–32	Light
14	60-slot	50–66	Heavy

Number of cases	Job type classification
48	Pure storage
13	Pure retrieval
29	Mixed storage and retrieval

Table 5.2 Performance gap for pure storage cases compared with the MILP model

Block dimension	Workload	Max (%)	Min (%)	Avg (%)
40-slot	Light	0.40	0.00	0.13
40-slot	Heavy	3.48	0.30	1.38
60-slot	Light	1.27	−0.16	0.26
60-slot	Heavy	4.24	0.36	1.37
Overall		4.24	−0.16	0.74

results provided by the MILP model. We adopt the total completion time (TCT) as used in Li et al. (2012) for the comparison purpose. Only pure storage test cases are compared; pure retrieval, and mixed storage and retrieval test cases are not comparable due to the fact that the weights assigned to retrieval earliness, retrieval lateness, and storage lateness between the two approaches are different.

It can be observed that the performance difference between the clustering-reassigning approach and the MILP model developed in Li et al. (2012) is very narrow. For light workload test cases in 40-slot blocks, the maximum performance gap is 0.40 % while the minimum performance gap is 0.0 % which indicates that the clustering-reassigning algorithm achieves the same results as the MILP model, the average performance gap is 0.13 %. The similar trend can be observed for the 60-slot blocks with light workload as well, with a maximum performance gap of 1.27 % and minimum −0.16 %, which implies that sometimes the clustering-reassigning approach provides slightly better results than the MILP model does. The average performance gap for light workload 60-slot test cases is 0.26 %. When the workload gets heavier, the clustering-reassigning algorithm's performance is moderately affected. For 40-slot test cases, the maximum performance gap is 3.48 % with an average gap of 1.38 %; for 60-slot test cases, the maximum performance gap is 4.24 % with an average gap of 1.37 %. For over all test cases, the average performance gap is 0.74 % which demonstrates that the clustering-reassigning approach provides satisfactory results compared to the MILP model.

In terms of solution time, on average the MILP model needs 2–3 s for light workload cases and 5–12 s for heavy workload cases, respectively (Li et al. 2012). Although the performance of the MILP model has been improved significantly compared to the discrete time model proposed in Li et al. (2009), the average time required indicates that the MILP model is still not very suitable for real-time scheduling, especially when many yard blocks are involved. Furthermore, to achieve the desired solutions, a mathematical optimization package, such as CPLEX must be required. The clustering-reassigning approach, on the other hand, takes very short time (i.e., instantly) in all cases due to the low algorithmic complexity, and thus makes the approach appropriate for real-time scheduling.

5.4 Conclusion

In this chapter, we discussed two basic underlying concepts for designing heuristic approaches for the single yard block multiple yard crane scheduling problem: concurrent scheduling and divide and conquer. We then describe a clustering-reassigning approach, which combines the advantages of both the concurrent and divide and conquer approaches. The proposed heuristic is efficient in dealing with both storage and retrieval container moves simultaneously. The yard crane interference, as well as other operational constraints, is realistically modeled and tackled using a novel conflict resolving mechanism. The clustering-reassigning approach provides solutions in polynomial time in terms of number of jobs to be scheduled and is therefore an ideal candidate for real-time yard crane scheduling. The proposed algorithm provides high quality results and the comparison with the optimization results generated by the continuous time MILP model confirms the algorithm's performance. Simulation runs, which will be discussed in Chap. 6, further validate the robustness and stability of the algorithm.

References

Bish, E.K., Chen, F., Leong, T., Li, C., Ng, J.W.C., Simchi-Levi, D. (2001) Analysis of a new scheduling and location problem. Naval Research Logistics. 46:363–385.

Cao, Z., D. H. Lee, Q. Meng. (2008) Deployment strategies of double-rail-mounted gantry crane systems for loading outbound containers in container terminals. International Journal of Production Economics. 115(1):221–228.

Dorndorf, U., F. Schneider. (2010) Scheduling automated triple cross-over stacking cranes in a container yard. OR Spectrum. 32(3):617–632.

Li, W., Goh, M., Wu, Y., Petering, M. E. H., de Souza, R., and Wu, Y. C. (2012) A continuous time model for multiple yard crane scheduling with last minute job arrivals. International Journal of Production Economics. 136(2):332–343.

Li, W., Wu, Y., Petering, M. E. H., Goh, M., and de Souza, R. (2009) Discrete time model and algorithms for container yard crane scheduling. European Journal of Operational Research. 198 (1):165–172.

Sharif, O., Huynh, N. (2012) Yard crane scheduling at container terminals: A comparative study of centralized and decentralized approaches. Maritime Economics & Logistics. 14(2):139–161.

Tou, J. T., R. C. Gonzalez. (1974) Pattern Recognition Principles. Addison-Wesley, Reading.

Wu, Y., Petering, M. E. H., Li, W.-K., Goh, M., and de Souza, R. (2006) A clustering-reassigning approach for multiple yard cranes scheduling at container terminals. In Pawar, K. S., Yu, M., Zhao, X., and Lalwani, C. S., editors, Proceedings of the 11th International Symposium on Logistics, pages 637–643, Beijing China.

Chapter 6
Validation via Simulation

Yard crane scheduling is just one part of a container terminal's operations management. Inevitably, yard cranes need to work and coordinate with other pieces of equipment. Some of them have direct interactions with yard cranes, such as prime movers (container trucks), others have indirect interactions with yard cranes, such as quay cranes. In order to assess the performance of any proposed yard crane scheduling solution, its standalone performance and the overall system performance with the solution integrated should be considered. For complex systems, computer simulation is a frequently used tool. Computer simulation refers to "methods for studying a wide variety of models of real-world systems by numerical evaluation using software designed to imitate the system's operations or characteristics, often over time" (Kelton et al. 2003, p. 7). Practically, simulation involves the process of designing and creating a computerized model so that numerical experiments can be conducted in order to better understand the system at hand (Kelton et al. 2003, p. 7). In this chapter, we use computer simulation to validate and assess the proposed heuristic algorithm in Chap. 5. We start with a simple simulation model which assesses the heuristic algorithm's performance at the yard block level. We then deploy a simulation model which models a mega transshipment container terminal and investigate the performance of the heuristic solution in Chap. 5, along with some other heuristics.

6.1 Simulation at the Block Level

In both the mixed integer linear programming (MILP) model in Chap. 4 and the clustering-reassigning approach in Chap. 5, although the YC travel time is modeled realistically, the handling time of each job is assumed to be fixed. In order to provide feasible solutions for practical usage, the handling time for each job is assumed to be a rather conservative constant amount of time when scheduling. While this can somehow guarantee that the schedules can be followed in practice,

W. Li et al., *Planning and Scheduling for Maritime Container Yards*,
DOI 10.1007/978-3-319-17025-1_6

the make-span, which corresponds to the duration from the start of the planning horizon to the finish of the very last job, derived from the schedules generated might give misleading information.

Therefore, it is desirable to see how well the solutions perform and how robust they are if these schedules are strictly followed by crane operators. In order to achieve this goal, simulation runs at the yard block level are conducted to validate the performance of the clustering-reassigning approach. By "strictly following," we mean that each job is handled strictly according to the sequence generated by the clustering-reassigning algorithm, but the handling time is assumed to follow a known triangular statistical distribution with parameters of (1.2, 2.0, 3.4) from a terminal operator.

The schedules generated by the clustering-reassigning approach are tested by simulating the handling times and each case is checked for 10 runs. Table 6.1 shows the simulated yard truck total waiting time compared to the scheduled results for pure storage cases. The items "Max," "Min," and "Avg" in the table stand for the maximum, minimum, and average percentage of the simulated runs compared with the scheduled results. It can be observed from the table that the maximum percentages of total waiting time in the light workload cases (both 40-slot and 60-slot) are equal to the scheduled ones, while for the other two heavy workload batches of test cases, the percentages are 76.83 % and 85.98 %, respectively. A similar pattern can be examined for the average percentage: in the light workload cases, the average percentages are closer to scheduled ones; while in the heavy workload cases, the deviation from the scheduled ones is large. This is due to the fact that under light workloads, the total waiting time is quite small even when using a conservative handling time for scheduling since the generated schedules are quite "loose." However, for heavy workload cases, the schedules are normally "tight;" therefore, some jobs are unnecessarily postponed due to the conservative estimation of the handling time, which in turn explains the reduction of total waiting time in simulation.

Table 6.2 lists the scheduled and simulated make-spans compared to the ideal ones. The ideal make-span is achieved by assuming that the last job in the job set can be handled on time. The scheduled and simulated make-spans are then compared with the ideal ones. Examining the "Scheduled" column in Table 6.2, it is found that for heavy workload cases the maximum percentages of scheduled results exceed the ideal ones by about 10−15 %. This indicates that in heavy workload yards, sometimes delay is inevitable. It can also be noticed that for light workload 40-slot cases, the maximum percentage is also high. This may due to the fact that in

Table 6.1 Total yard truck waiting time in simulated runs versus schedules

Block dimension	Workload	Max (%)	Min (%)	Avg (%)
40-slot	Light	100	11.92	77.72
40-slot	Heavy	76.83	42.27	58.35
60-slot	Light	100	43.40	80.91
60-slot	Heavy	85.98	40.16	58.35

Table 6.2 Scheduled and simulated make-spans versus ideal ones for pure storage cases

Block dimension	Workload	Scheduled (%)			Simulated (%)		
		Max	Min	Avg	Max	Min	Avg
40-slot	Light	114.71	100	101.7	114.01	100	101.42
40-slot	Heavy	110.11	100	103.06	106.06	100	101.57
60-slot	Light	100.73	100	100.14	100.64	100	100.02
60-slot	Heavy	113.92	101.47	105.42	111.8	100	103.72

40-slot yards, the chance of crane interference will be higher than in 60-slot cases. The minimum percentage is somewhat equal to or close to 100 %, which indicates that the jobs can be finished within the ideal make-span for some cases. Average percentage of scheduled results indicates that the make-spans are very close to the ideal ones in light workload cases and slightly bigger in heavy workload cases. The simulated results show the same trend as the scheduled ones with the maximum and minimum percentages bounded by the scheduled results. The average results indicate that the algorithm can generate schedules which enable the work to be handled with a slight delay in make-span for both light and heavy workloads but still bounded by the scheduled results. This will ensure that the planning and deployment of yard cranes can rely on the schedules generated.

In all test cases, the simulated total waiting time is less than the scheduled one, and the make-span always lies in the range generated by the clustering-reassigning algorithm; leading us to conclude that the clustering-reassigning approach provides a reasonable solution to the multiple yard crane scheduling problem and the results can be used to guide real-life operations. In addition, the algorithm's ability to providing instant solutions is one of the essential features for real-time scheduling. Furthermore, the proposed algorithm might be very useful in assessing the workload for different yard blocks and determine the inter-block yard crane movements.

6.2 Simulation at the Terminal Level

Simulation at the terminal level provides a holistic view of the container terminal operations. It also allows the examination of impacts of different solutions to the overall terminal performance. A discrete event simulation model, which tracks the detailed movement of each container passing through a container terminal, is developed to gauge the merit of the various systems. The simulation model uses a modular structure which enables the "plug and play" of different yard crane scheduling algorithms. The main output from the simulation model is the average, long-run quay crane rate (i.e., gross crane rate, GCR) that a given combination of yard crane scheduling, yard crane routing, and prime mover dispatching algorithms can achieve. This section is based on the report from Petering et al. (2006), and complete description of the simulation model can be found in Petering (2007), Petering and Murty (2009), and Petering et al. (2009).

6.2.1 The Simulation Model

The simulation model accommodates both 20′ and 40′ containers. The prime movers can carry up to two 20′ containers on a single level. All containers are vessel-to-vessel transshipment and therefore, gate operations are not considered. The event-scheduling/time advance algorithm is used to make sure that time advances properly and all events occur in the correct chronological order. The structure that drives the simulation program is the future event sequence (FES). It is always sorted by increasing event time; thus, the first event on this list is the next event that occurs. Overall, there are two kinds of events: primary and secondary. Only primary events may be listed in the FES. A primary event generally (a) causes an immediate change in the state of the container terminal, (b) causes other events, some primary and some secondary, to occur immediately along with the primary event, and (c) causes other primary events to be placed at some future time in the FES. Table 6.3 shows the

Table 6.3 Events in the simulation program

Type	Event #	Event description
Primary events	0	Check terminal status consistency
	17	Vessel arrives
	22	Vessel finishes berthing
	16	Vessel finishes unberthing
	13	All QC job sequences scanned
	24	QC finishes handling
	11	YC finishes cross gantry
	19	YC finishes linear gantry
	23	YC finishes handling
	25	PM finishes journey
Secondary events	15	Vessel starts berthing
	2	Vessel starts unberthing
	14	QC job sequences generated
	1	QC job sequences erased
	21	QC starts handling
	20	Online storage algorithm called
	12	General retrieval and storage algorithm called
	10	YC cross-gantry scoring module called
	9	Multiple YC scheduling algorithm called
	5	Individual YC routing algorithm called
	5A	Individual YC routing on-the-fly algorithm called
	3	YC starts cross gantry
	4	YC starts linear gantry
	18	YC starts handling
	8	Workcenter PM assignment algorithm called
	7	Individual PM dispatching algorithm called
	6	PM starts journey

primary and secondary events that occur within the simulation model. Overall, there are 10 types of primary events that can be placed in the FES. A primary event can instantly lead to one or more primary events, or one or more of the 17 secondary events, which in turn can instantly lead to more primary or secondary events.

In the simulation program, primary events often immediately trigger other primary and secondary events, and cause new events to be placed in the FES. Figure 6.1 shows the program flow when a new event in the FES is processed. Each solid rectangle represents an event that is unfolding during the running of the simulation program.

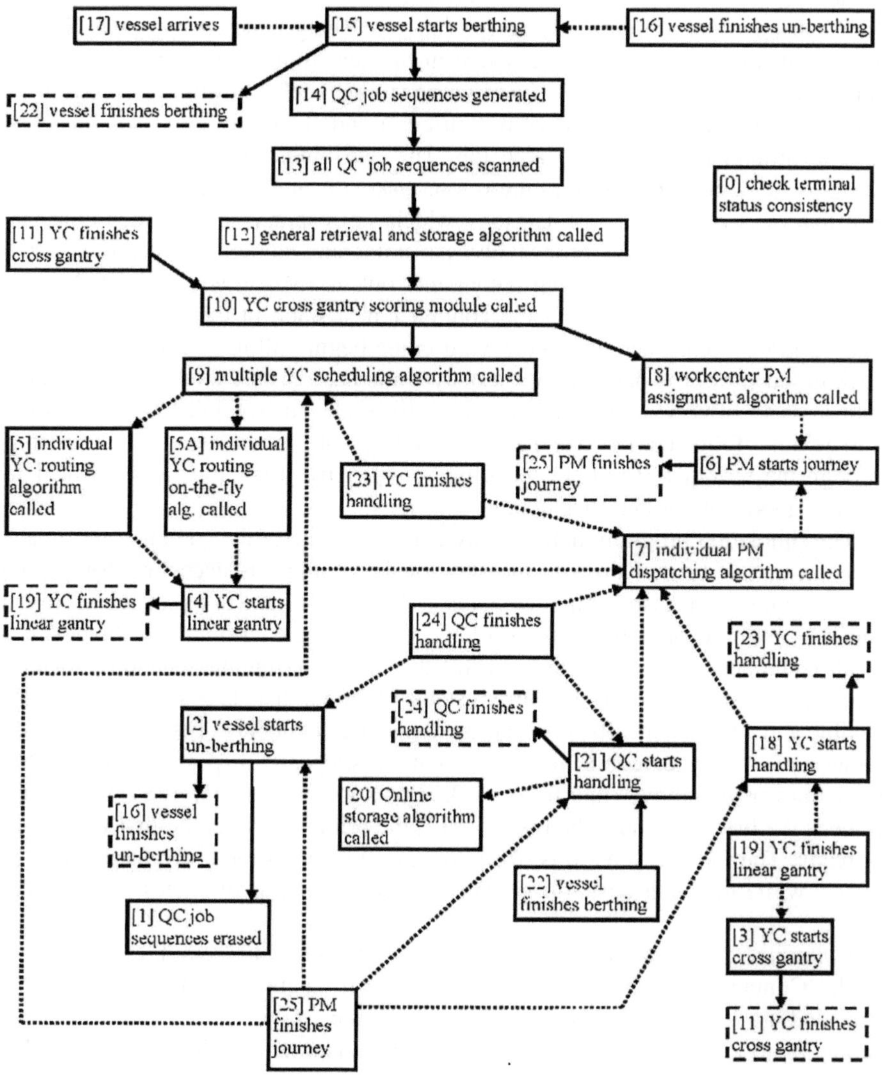

Fig. 6.1 Processing a new event in the FES

A dotted rectangle indicates that an event is being placed in the FES. Solid arrows indicate a guaranteed immediate and instantaneous flow from one event to another. Dotted arrows indicate that a call to another routine (i.e., the immediate triggering of another event) is possible, but not guaranteed. If the events connected by such arrows both refer to physical processes undertaken by a machine, usually a "handshake" criterion must be met for the triggering to occur, which usually results in a container being transferred between two machines in this case.

The concept of a *job* is central to the running of the simulation program. A job is defined as the movement of a single container (either 20′ or 40′) from yard to quay or vice versa. Throughout each simulation run, the program carefully maintains a list of all jobs that are currently active (i.e., are being handled by some piece of equipment) or are expected to be active in the near future. The capital letters 'U' (quay crane unloading) and 'S' (yard crane storage) refer to different tasks in the same job. Similarly, 'R' (yard crane retrieval) and 'L' (quay crane loading) also refer to different tasks in the same job. The two kinds of job in the simulation program are named according to these four tasks, i.e., unloading-storage (U–S) job and the retrieval-loading (R–L) job, which correspond to the incoming and outgoing container flows, respectively. An unloading-storage job becomes active when a quay crane begins lifting a container out of a vessel and is deleted once a yard crane has finished placing the container in a stack in the yard. A retrieval-loading job becomes active when a yard crane begins lifting a container out of a stack in the yard; it is deleted once a quay crane has finished loading the container onto a vessel. Every container eventually gives rise to two jobs: one unloading-storage job, followed by one retrieval-loading job as the container transships through the terminal. Tracking the status of all jobs in the job list is critical for making good routing and dispatching decisions.

The simulation program meticulously tracks the state of the container terminal at every instant of time. Knowing the state of the terminal is necessary for making good routing and dispatching decisions. The real-time status of the terminal depends on (a) the operational status of every entity at the terminal, (b) the positions of all containers in the storage yard, (c) the up-to-the-minute schedules for each piece of equipment, and (d) the current status of the job list. This state framework is flexible and can realistically simulate the transfer of containers between cranes and prime movers. For example, if a quay crane is supposed to pass a container to a prime mover, the above state framework allows for two possibilities: (1) the quay crane is the first machine to be ready, or (2) the prime mover is the first machine to be ready. The former scenario is likely to happen when prime movers are relatively scarce; while the latter scenario is likely to occur when quay cranes are relatively scarce.

Container groups are used to allocate locations for containers in the simulation model. Containers in the same group have the same length, weight class, destination port, and are loaded onto ships from the same liner service. Therefore, containers belonging to the same group can be considered indistinguishable for vessel loading purposes. In the container yard, we require that all containers in a stack belong to the same group. In other words, the contents in each stack are homogenous. This means that there is no need to shift containers from one stack to another in the yard in order to retrieve one particular container. In the simulation program, the sequence of groups that a quay crane handles is fixed and cannot be changed. However, each quay crane may handle the individual containers within a particular group in any sequence it chooses. Hence, during vessel loading, a quay crane may serve whichever prime mover that first arrives with a container of the appropriate group.

Prime movers transport containers between the container yard and the quay. In the simulation program, prime movers receive relatively specific instructions. In particular, prime movers are always instructed to travel to either a specific ground slot in the yard or a specific quay crane. They do not travel to general areas like "block 2" or "berth 7." Once a prime mover arrives at its destination, it may only pick up or drop off a container. Each prime mover sees only one instruction at a time, which tells the prime mover to proceed to a specific location and wait there until either a yard crane or a quay crane chooses to initiate a container transfer with the prime mover at that location.

Yard cranes are responsible for transferring containers between prime movers and stacks in the yard. Like prime movers, yard cranes receive relatively specific instructions, i.e., yard cranes are always instructed to travel to a specific ground slot in the yard. The first part of a yard crane's instruction is always a linear gantry move. Once a yard crane finishes the linear gantry and arrives at the destination ground slot, it will have further instructions to (1) make a cross-gantry move, (2) handle a container at that ground slot, or (3) become idle. For case (1), the yard crane's destination ground slot is always either the first or last ground slot in a block. For case (2), the yard crane may only handle a container that belongs to the group listed next in its job sequence. In addition, the container transfer must be of a specific type: 'S' (pick up from a prime mover and place in a stack) or 'R' (retrieve from a stack and give to a prime mover). If there is no particular prime mover specified, the yard crane may serve any prime mover that is at the destination ground slot and has the correct status. If no such prime mover is present and the job type is 'S', the yard crane will wait at the ground slot until a prime mover with instructions containing the appropriate group and job type arrives at that ground slot. Once the prime mover arrives, the yard crane begins the handling operation. However, when the job type is 'R' and there is no prime mover present, the yard crane will go ahead with the retrieval and a prime mover with the appropriate status might arrive in the interim. The yard crane might need to wait if no prime mover arrives after it has finished the retrieving task. Case (3) is self-explanatory.

The quay cranes receive similar instructions as the yard cranes. The difference is that quay cranes are considered immobile and therefore their gantrying is not modeled. It should be noticed that although a detailed sequence of future jobs is maintained for any quay crane, each quay crane sees only one instruction at a time. Examples of quay crane instructions are "L, group 319" and "U, group 9119". The sequence of future instructions for each QC is fixed for a look-ahead horizon of a certain number of minutes.

In order for a container to pass from one machine (i.e., yard crane, quay crane, or prime mover) to another machine, the status of both machines must meet a "handshake" protocol. The handshake protocol requires that: (1) both machines must be at the same location (slot or quay crane); (2) both machines must have instructions to be at the same specific location (ground slot or quay crane); (3) both machines must have instructions to perform the same type of transfer ('U', 'S', 'R', or 'L'); and (4) both machines must have instructions to interact with a container in the same group. If these four criteria are met, a container of the appropriate group will be transferred from one machine to the other.

6.2.2 The Results

The simulation program was compiled using Microsoft Visual C++ 6.0. Experiments were run using four different desktop computers with CPU speeds ranging from 2.80 to 3.06 GHz and RAM ranging from 504 MB to 2 GB. Table 6.4 gives the major specifications of the container terminal we consider. All vessels calling at the terminal are assumed to have the same size and unload the same number of containers. Figure 6.2 depicts the physical layout of the terminal. As Fig. 6.2 indicates, prime movers may travel two ways on all vertical lanes and on the two horizontal lanes at the top and bottom of the terminal, but may only travel one way on the horizontal lanes in the middle of the terminal.

The simulation terminates after 270 vessels are fully processed at the terminal. Vessels are not allowed to leave the terminal unless all of their boxes have been unloaded and loaded. As explained earlier, a yard control system has many subsystems and therefore, each combination of these subsystems is the result of combining algorithms for handling the tasks of (1) interzone yard crane deployment, (2) yard crane routing and scheduling within each zone, and (3) prime mover dispatching.

We used the simulation model developed in Petering et al. (2006) and tested the performance of the clustering-reassigning approach in Chap. 5, together with some rule of thumb approaches that are commonly used in terminal operations by crane operators. The crane scheduling approaches considered are summarized in Table 6.5. They include 10 approaches based on rules of thumb and one heuristic approach—the clustering-reassigning algorithm described in Chap. 5.

Table 6.6 presents the performance of different yard crane routing policies under the settings that yard crane/quay crane ratio is 2.5, prime mover/quay crane ratio is 7, and no cross gantrying and no prime mover dual loading are allowed. Two

Table 6.4 Container terminal specifications for experiments

Type	Quantity
Berths	9
Vessel calls per week	90
Expected quay crane lifts per week	172,800
Yard blocks	90
Quay cranes	36

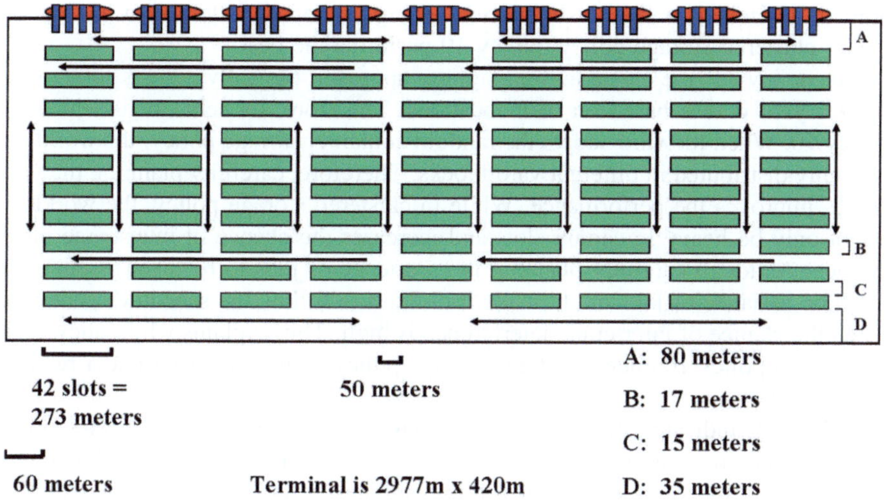

Fig. 6.2 Terminal layout in the experiments (Petering et al. 2006)

Table 6.5 Yard crane routing approaches considered in the simulation

Policy	Description
1	Earliest arrived prime mover
2	Nearest container
3	Earliest 'R' task prime mover; otherwise earliest 'S' task prime mover
4	Earliest 'S' task prime mover; otherwise earliest 'R' task prime mover
5	Nearest 'R' task; otherwise nearest 'S' task
6	Nearest 'S' task; otherwise nearest 'R' task
7	Most urgent job
8	Most urgent 'R' task; otherwise most urgent 'S' task
9	Most urgent 'S' task; otherwise most urgent 'R' task
10	Most urgent 'R' task; otherwise nearest 'S' task
11	Multiple yard crane scheduling approach developed in Chap. 5

Table 6.6 Performance of the approaches considered

Storage	1	2	3	4	5	6	7	8	9	10	11
Cluster	25.4	26.6	26.5	23.8	27.2	24.3	23.5	26.8	23.8	27.7	26.8
Random	26.4	28.2	26.7	24.7	28.6	25.7	25.2	27.1	25.3	28.9	29.8

storage policies are used: storage based on clusters, and random storage (Petering et al. 2006). It can be observed that the performance of each yard crane routing policy under random storage outperforms the respective one under clustered storage policy. The best yard crane routing policy under random storage policy is policy number 11, which is the clustering-reassigning approach, followed by policy 10. Under clustered storage policy, policy 10 performs the best.

For policy 11, the clustering-reassigning approach basically assigns jobs to the nearest yard cranes based on their locations. Therefore, its behavior is similar to policy 10. When the yard adopts the random container storage policy, the workload is evenly distributed in different yard blocks. Therefore, carefully planning the yard cranes improves the performance. While in clustered storage policy, the workload might only be heavy in certain slots and increases the chance of inter-crane interference. Although carefully planning the sequence of jobs to be taken is possible, the time variance of jobs might make it difficult to follow the schedules generated when the chance of inter-crane interference is high. This explains why policy 11 is worse than policy 10 under clustered storage policy while it is better under random storage policy.

Table 6.6 indicates that when yard crane/quay crane ratio is 2.5 and prime mover/yard crane ratio is 7, sometimes the clustering-reassigning approach is not as competent as the rules of thumb routing policies 5 and 10, although overall it provides a decent performance. The power of the clustering-reassigning approach unveils when the number of pieces of equipment is large. Table 6.7 shows the performance of yard crane routing policy 11 compared to policy 10 (which performs best among all rules of thumb routing policies) under random storage and clustered storage policies when no cross gantrying and no prime mover dual loading are allowed. For both policies, the random storage policy is always better than the clustered storage policy, except for policy 10 when the YC/QC ratio is 2 and PM/QC ratio is 5. In general, with the clustered storage policy, the performance of policy 10 is better when the YC/QC ratios are 2.0 and 2.5; but policy 11 is better than policy 10 when the ratios are 3.0 and 3.5. When the random storage policy is adopted, the clustering-reassigning approach is always better than policy 10. This indicates that when the yard is not very "crowded," good rule of thumb routing policies would be sufficient to give instructions to different pieces of equipment; while in a more "crowded" environment, carefully scheduling the yard cranes is beneficial for the whole terminal's performance. It should also be noted that, when

Table 6.7 Performance of routing policies 10 and11 under different equipment ratios

Average QC rate (lifts/h)		Prime mover/quay crane					
		YC policy 11			YC policy 10		
Storage	YC/QC	5	7	9	5	7	9
Cluster	2	21.3	24.3	26.2	21.6	24.9	27.3
	2.5	24.4	26.8	28.4	24.1	27.7	29.6
	3	25.7	28.1	29.8	24.7	28.4	30.1
	3.5	26.3	28.9	30.1	24.7	27.6	29.1
Random	2	22.9	26.2	28	21.2	24.9	27.5
	2.5	26.5	29.8	32.4	24.6	28.9	31.8
	3	28	31.9	34.1	26.3	30.7	33.3
	3.5	28.9	32.6	34.7	26.7	30.8	32.8

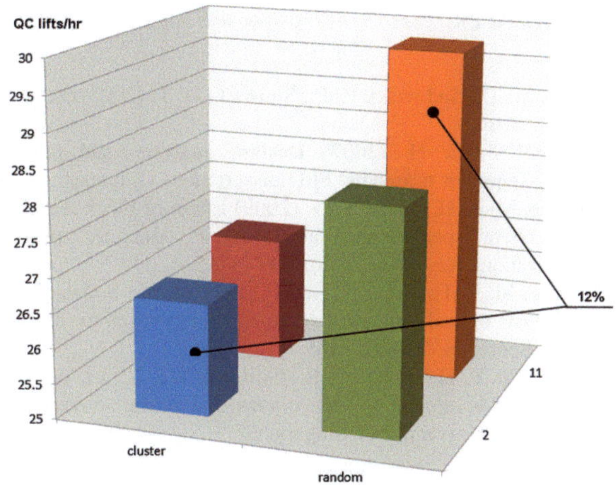

Fig. 6.3 GCR improvement compared to baseline using policy 11

using heuristic policies, the performance of YC/QC ratio 3.5 is normally worse than YC/QC ratio 3.0 due to the lack of inter-crane interference handling mechanism. However, when using the clustering-reassigning approach (policy 11), the performance of YC/QC ratio of 3.5 is always better which demonstrates that the heuristic approach can handle inter-crane interference well.

From the above analysis, it is shown that the yard crane routing policy can generate big impacts on the container terminal overall system performance. As Fig. 6.3 shows, under the settings of YC/QC ratio of 2.5 and PM/QC ratio of 7, the combined effect of simply changing the yard crane routing policy and storage policy will be a massive 12 % performance improvement. Furthermore, if the ratio of yard crane/quay crane is increased from 2.5 to 3.0 and prime mover/quay crane ratio is changed from 7 to 9, the improvement will be even larger.

6.3 Conclusion

This chapter validated the performance of the proposed yard crane scheduling algorithm in Chap. 5. Both simulations at the block level and at the terminal level were conducted. The results indicated that the clustering-reassigning approach can generate high-quality solutions in very short time and it works well with other pieces of equipment in the simulation model. It can greatly improve the overall terminal performance against some commonly deployed rule of thumb approaches, and the performance improvement will be more conspicuous when the random storage policy is employed at the terminal.

References

Kelton, W. D., Sadowski, R. P., Sturrock, D. T. (2003) Simulation with Arena (3rd edn). McGraw Hill, Boston.

Petering, M. E. H. (2007) Design, analysis, and real-time control of seaport container transshipment terminals. PhD dissertation, University of Michigan, Ann Arbor, Michigan.

Petering, M. E. H., Murty K.G. (2009) Effect of block length and yard crane deployment systems on overall performance at a seaport container transshipment terminal. Computers and Operations Research. 36(5):1711–1725.

Petering, M. E. H., Wu, Y., Li, W., Goh, M., de Souza, R. (2009) Development and simulation analysis of real-time yard crane control systems for seaport container transshipment terminal. OR Spectrum. 31: 801–835.

Petering, M. E. H., de Souza, R., Goh, M., Wu, Y., Li, W., Loy, H. T., Ku, L. P., and Yow, K. Y. (2006) Movement of yard cranes and prime movers at PSA Singapore. Technical report, National University of Singapore.

Index

B
Berth allocation problem (BAP), 16, 17, 24
Bi-index, 42, 70

C
Clustering-reassigning
  clustering, 88, 89, 91–94, 97, 99, 104, 106
  reassigning, 88, 89, 91–94, 97, 98, 104, 106, 107
Complexity, 15, 17, 19, 24, 85, 88, 91, 94
Concurrent scheduling, 85–88, 94
Container yard (CY), 1, 6, 31, 54, 102, 103
Continuous time, 23, 56–58, 68, 70, 92
CRHA, 73–75, 78, 81
Cross gantry, 21, 33, 54, 59, 104, 106

D
Discrete event simulation, 21, 55, 99
Discrete time, 56, 57
Divide and conquer, 85, 87, 88, 94

F
FEU, 5

G
Gross Crane Rate (GCR), 10, 33, 99, 107

I
Intercrane interference, 22, 34, 35, 38, 40, 54–56, 65, 66, 88, 89, 106, 107

J
Job handling range (JHR), 42, 43, 70, 71
Job handling time (JHT), 36, 44, 46, 67, 76, 77

L
Last minute job, 76

M
Mixed Integer Linear Programming (MILP), 3, 19, 35, 40, 41, 48, 56–58, 67, 73, 75, 76, 91–93, 97

N
Numerical experiment, 91, 97

P
PM waiting time (PMWT), 41–43, 48, 57
PSA Singapore, 6, 7, 9–14

Q
Quay crane assignment (QCA), 16, 17, 19
Quay crane scheduling (QCS), 16–19, 55

R
Retrieval earliness (RE), 36, 68, 86, 88, 89
Retrieval lateness (RL), 38, 46, 66, 89, 93
Rolling horizon algorithm (RHA), 44, 46, 72
Routing policy, 105–107
Rubber Tired Gantry Crane (RTGC), 13
Rule of thumb approaches, 104

S
Simultaneous storage/retrieval, 23, 35, 44, 54
Storage lateness (STL), 38, 47, 66, 86, 88, 89, 93

T
Target time, 21, 35–37, 40, 42, 44, 46, 65, 66, 68, 70–73, 89
TEU, 5, 13, 53
Tri-index, 41, 42, 70

V
Validation, 97, 98

W. Li et al., *Planning and Scheduling for Maritime Container Yards*,
DOI 10.1007/978-3-319-17025-1

W
Work list, 21, 33, 66

Y
Yard crane scheduling (YCS), 20, 22, 24, 31, 53, 99, 85
YC separation distances, 31, 53, 56

MIX
Papier aus verantwortungsvollen Quellen
Paper from responsible sources
FSC® C105338

If you have any concerns about our products,
you can contact us on
ProductSafety@springernature.com

In case Publisher is established outside the EU,
the EU authorized representative is:
Springer Nature Customer Service Center GmbH
Europaplatz 3, 69115 Heidelberg, Germany

Printed by Libri Plureos GmbH
in Hamburg, Germany